THE
Z
FACTOR

THE

Z

FACTOR

**MY JOURNEY AS THE
WRONG MAN
AT THE
RIGHT TIME**

SUBHASH
CHANDRA

with PRANJAL SHARMA

HarperCollins *Publishers* India

First published in India in 2016 by
HarperCollins *Publishers* India

Copyright © Subhash Chandra 2016

P-ISBN: 978-93-5177-324-5
E-ISBN: 978-93-5177-325-2

2 4 6 8 10 9 7 5 3

Subhash Chandra asserts the moral right to be identified
as the author of this work.

The views and opinions expressed in this book are the author's own and
the facts are as reported by him, and the publishers
are not in any way liable for the same.

HarperCollins *Publishers*

A-75, Sector 57, Noida, Uttar Pradesh 201301, India
1 London Bridge Street, London SE1 9GF, United Kingdom
Hazelton Lanes, 55 Avenue Road, Suite 2900, Toronto, Ontario M5R 3L2
and 1995 Markham Road, Scarborough, Ontario M1B 5M8, Canada
25 Ryde Road, Pymble, Sydney, NSW 2073, Australia
195 Broadway, New York, NY 10007, USA

Typeset in 12/14 Warnock Pro Light at
SÜRYA, New Delhi

Printed and bound at
Thomson Press (India) Ltd.

CONTENTS

Acknowledgements   *ix*
Preface   *xi*
Family Tree   *xiv*

1. SMALL BEGINNINGS   1
   Three brothers set up a market

2. A CRISIS IN THE FAMILY   19
   I start work as a teen to pay off debt

3. THINKING OUT OF THE BOX   31
   Convincing FCI to adopt a new process

4. YOUNG MAN ON THE MAKE   39
   Living off bluff, bluster and gumption

5. BETWEEN THE SHEETS   50
   Or how my hair turned grey overnight

6. DIVERSIFYING INTO PACKAGING   61
   Acchey din arrive as I spread my wings

7. POLE-VAULTING TO A HIGHER PLANE   65
   I take a risk but get cheated...again!

8. THE RUSSIANS ARE COMING   75
   A Swamiji, the Gandhis, and a very big deal

9. OF RICE...AND AVARICE   87
    The Soviets smell something fishy about basmati!

10. A LATE-NIGHT MEETING WITH MRS G   97
    Feeling like an ant in a fight between elephants

11. GOOD THINGS COME IN SMALL PACKAGES   105
    After teething troubles, we hit pay dirt

12. A ROUGH ROLLER-COASTER RIDE   113
    Some people in government are not amused

13. BROADCASTING MY INTENTIONS...   119
    ...but no one takes me seriously

14. THE $5 MILLION GAMBLE   126
    Despite an incredible offer, I am still the last choice!

15. IT'S SHOWTIME, FOLKS!   138
    A quiet launch but a new approach to producing shows

16. WHO WILL MANAGE THE MANAGERS   154
    Renewing teams to keep up the pace

17. THE NEWSROOM   161
    A change in my profile as I get into current affairs

18. A ROCKY PARTNERSHIP   175
    Dealing with Mr Murdoch—ally and rival both

19. SETTING MY HOUSE IN ORDER   191
    A family division...and a Tehelka *sting*

20. FRIENDS AND ENEMIES   201
    Taking on the Ambanis...reluctantly

21. LOSING $6 BILLION...   206
    ...as Zee's stock skyrockets, then crashes

22. IT'S NOT CRICKET   211
    TV rights...and wrongs...and a controlling board

23. CALMING INFLUENCES   222
    Vipassana brings peace...but turmoil, too

24. THE ART OF MANAGEMENT   237
    Staying focused on the present

25. A SATELLITE PROJECT RUNS AGROUND   247
    Losing big time in a high stakes game

26. STABBED IN THE BACK   257
    Steeling myself as the system turns against me

27. NEW HORIZONS   265
    Paving the road to success

    EPILOGUE   271

    Appendices   273
    Index   275
    About the Authors   281

ACKNOWLEDGEMENTS

I HAVE A strong belief in India's youth. I believe that out of every thousand kids, two are prodigies. Most prodigies don't get the opportunity to develop their talent and skills. Only a few are able to showcase them. These kids could be great entrepreneurs or professionals in the corporate world. They could join the government, politics, media or judiciary. But despite the lack of chances, they face the world stoically and smile in adversity.

This book is not only dedicated to the youth but to all those who live with a smile despite the constraints. My friends from abroad often ask me how so many poor Indians smile despite their difficult lives. Why aren't they angry and bitter? I tell my foreign friends that Indians know how to prioritize their lives in pursuit of happiness, that they focus on what they have and not just on what they don't.

I would like to acknowledge my wife Sushila, who has put up with me most of my life. I have been guilty of not spending enough time with her. Even this book cut into my time with her. I would like to thank my brother Jawahar and my colleagues Himanshu Mody and C.S. Vishwanathan for assisting me with the facts for this book.

I thank Pranjal Sharma for having understood the complexities of my life and for telling my story in the right words, which were not far from mine. I thank my editor Amit Agarwal and publishers HarperCollins India for giving shape to the book.

I hope the book will not just be a good read but also help the readers overcome the difficulties in their lives. I hope they begin to believe in themselves, in their strength and capabilities.

PREFACE

WE INDIANS TEND to be docile. We prefer to be led and not be leaders. My working life began with this belief. I was content to be a follower. This changed at a very young age when my family faced a financial crisis. Our modest business in the small town of Hisar in Haryana was almost wiped out. I was shocked to see my grandfather and guru, Jagan Nath Goenka, a broken man. He was my idol, my inspiration. I saw him as a man of iron, someone who was infallible, who could do no wrong. He was the patriarch who supported an extended hundred-member family. How could he be so weak and helpless?

I asked myself, would I also be like that? Would I, too, accept the fate that had befallen him and the family?

I was only seventeen and faced a bleak future. But a voice inside me refused to accept the situation. I resolved to fight the circumstances. I would strive to revive my family's fortunes. Since then, I have never taken 'no' for an answer.

I wonder what people close to me think about this attitude. My sons, my colleagues, must wonder why I never accept a 'no'.

Do I worry when they question my stubborn behaviour? Not really. When I make up my mind, I don't give up.

Some people may think I am too old to have this attitude. But I do not feel the age, as I enjoy what I do. I feel responsible towards not only my family but also the society at large. I feel a responsibility towards humanity. I believe in 'Vasudhaiva

Kutumbakam'—the whole world is one family. My thinking is far more spiritual now than in my early years.

Many business leaders feel constrained by the investment climate in India, which still has so many regulations and restrictions. They feel the need to curry favour with officials and ministers in Lutyens' Delhi.

Some of these business leaders are too timid. Three of the projects that I began were ahead of policy regulations. I started a broadcasting business, a satellite communication project, and an amusement park, much before others. The government's understanding of these businesses was behind the times. I did not have the patience to wait for the government to understand these new sectors. I can say that I broke the law by starting a private television channel when it was not allowed. But I can also say that I forced the government to realize the importance of allowing private sector broadcasting to grow in India.

My thinking still is that if you believe in your idea and your intentions are honourable, you will not be stopped. At worst, people and the system may succeed in merely halting your progress.

My life also taught me self-belief. I have learnt to have faith in myself and my capabilities. When you have the passion and commitment, nature helps you succeed. It provides you a motivation to keep trying. There is no defeat or failure unless you have accepted it.

I launched a motivational talk show on Zee a few months ago. The *Dr Subhash Chandra Show* was launched as a result of a conversation I had with my father two years ago. He said, 'I know you are practising the family tradition of "Dasaund" (or donating at least 10 per cent of your profits) but have you shared your knowledge and experience with others to help them?' This set me thinking and I launched this show where I talk to young Indians and try and address their concerns about life's challenges. I am humbled that many of them say that my words have helped them.

I am troubled by certain developments in India. A few self-centric people are undermining our economy and society. A few greedy people hoard large amounts of wealth at the cost of other citizens. They use illegal means to earn money and exploit the system for their benefit. I am also concerned about the attempts to damage the fabric of our society. Nationalism is being maligned by people. Anyone who raises nationalistic issues is accused of being communal and divisive. I believe Indian society is strong and we will continue to live in harmony. My involvement in the Ekal School movement reflects my faith in India. I remain very optimistic about the strong and bright future of our nation.

6 December 2015   **Subhash Chandra**

FAMILY TREE

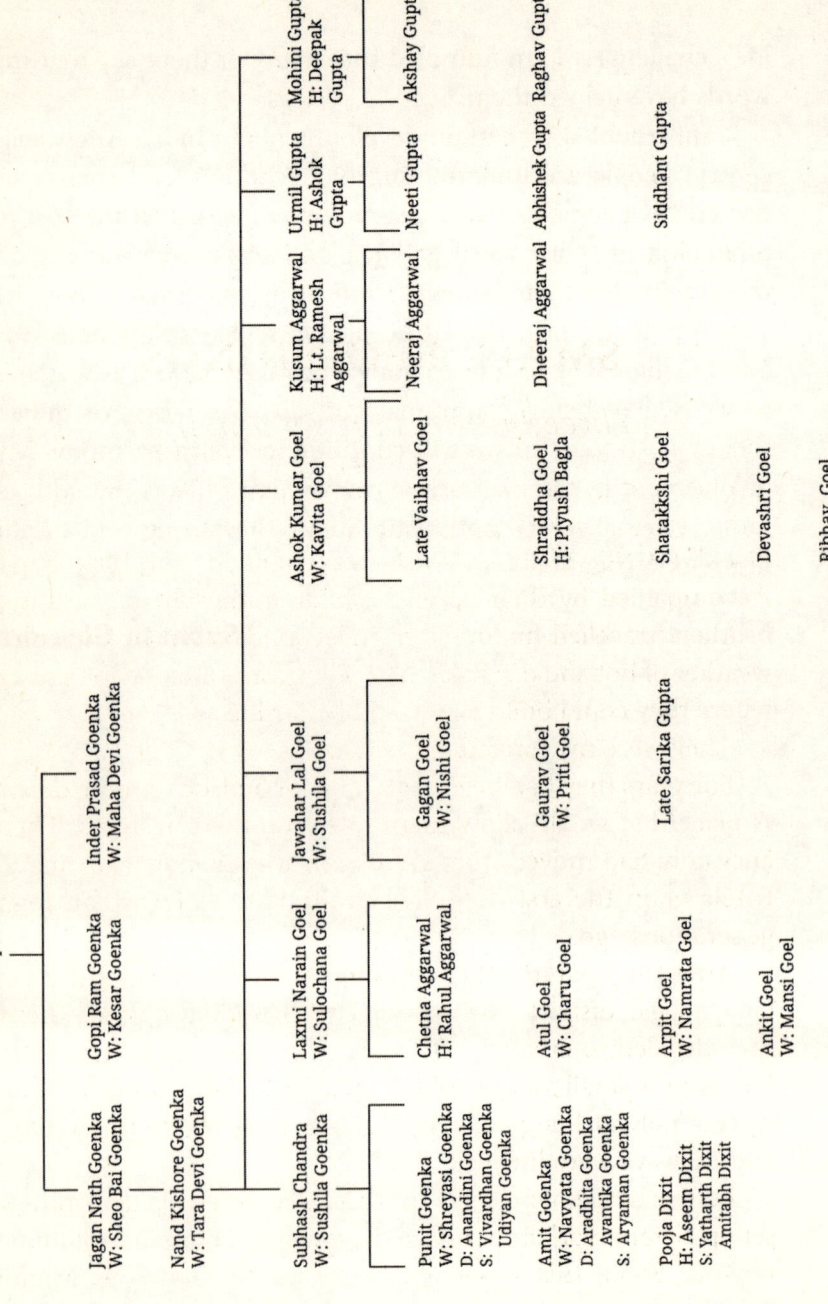

1

SMALL BEGINNINGS

Three brothers set up a market

EVERYTHING THEY OWNED was packed in cloth bags. Accompanied by their parents, children and wives, the three brothers travelled hundreds of miles, undaunted by the harsh weather of hot and dry Rajasthan. They came looking for a place where they could build a new and better life.

They were my forefathers.

For years they had been searching for a place to settle down. A place that would allow them to set up their business. Their ancestors had moved from Agroha in what is now Haryana to Fatehpur in the erstwhile Shekhavati state of Rajasthan forty generations ago.

After Fatehpur, the three brothers tried to settle in Bhadra in Ganganagar district. Not satisfied with trading options there, they decided to create a settlement just a few kilometres from Hisar, in a small village called Sadalpur. They chose the spot since a railway line passed by it. For people at the time, rivers and railways were lifelines.

At this settlement, in 1926, the three enterprising brothers set up a grain market to service the needs of Hisar and adjoining regions. Soon, this settlement came to be known as Mandi Adampur.

The town itself was little more than a strip of road, defined by rows of buildings on both sides. The main bazaar road was the centre of the town. It was just a 400-metre stretch but all the main shops, commercial establishments and residences were on this road. Business was transacted on the ground floor while families lived on the upper floors. This road was almost parallel to the railway line a few metres away.

As the grain market grew and developed into an important regional economic centre, the railway authorities decided to build a station for traders and farmers. This further helped Mandi Adampur grow into an agro-commerce centre.

The community came together to build basic amenities for themselves. They built a school that was later handed over to the government. The brothers also built an inn or *dharamshala* for travellers, an open-storage water tank, and a temple with their savings of Rs 20,000. These efforts met most of the needs of the trading community. The tiny settlement began to mature into a small town.

It was in this small town that I was born in 1950. By this time, the family business of trading had grown. The clan had three houses where the families of the three brothers lived.

My education began from age four in the local government school that had been originally built by the community. My early memories of Adampur are of a happy, simple life. Our town did not have electricity. All of us kids played in the dusty fields around our homes. The homes were connected on the first floor so that the ladies of the house could be in touch without having to come down to the market, which was dominated by men.

The men of the family slept outside in the outer courtyard, commonly called the *chabootra*, that extended from the steps of the shop. Or sometimes, in the summer, on the open roof of the house. Often there would be strong gusts of wind blowing in the night. And upon waking up we would find ourselves covered in sand. The area in and around Adampur was arid and dusty.

Going to a river or a canal was a treat for us. My father would take the children of the family and his friends for picnics to a river canal about a kilometre away. However, the women of the family would stay home and never joined in these outings.

I am told that right from my childhood, I was very clear about what I wanted. Once, when I was down with chicken pox, I demanded a car from my father. He asked me what type of 'toy car' I wanted. I said I wanted a real car, a Fiat. My father said, 'I will get one when I go to Hisar.'

'No, that will be too late. How much is a Fiat?' I asked. 'About Rs 16,000,' my father said.

'Okay, you place Rs 16,000 under my bed now. I will keep it with me until you leave to buy the car.'

My father actually put a few thousand rupees under my bed so that I could sleep. Without that promise, I would not have slept. Of course, by morning I had forgotten about it.

We had interesting visitors to our small settlement in Adampur. Every two or three months some itinerant fakirs and sadhus would visit the mandi. They followed sanatan dharma and were usually from religious towns like Rishikesh or Haridwar. At times, we would see travelling Jain monks, both male and female (sadhvis).

During the monsoon season, these travelling monks would stay put in one town for a couple of months. The rest of the year, they would not spend more than a couple of days in one place. I grew close to one of them, Sadhvi Gulaba Bai. I would attend all her discourses/lectures, which were held mostly in the evenings.

When she was alone, I would ask her questions that would surprise her, coming as they were from a precocious six-year-old. What is life? What happens after life? What is alive, what is dead? What is our purpose, what is human life? Why are we born humans and not animals? And so on. Gulaba Bai would indulge me and laughingly tell my father how I asked her too many tough questions. My parents, though, would encourage me to spend time with her.

I was hardly a saintly child with a halo, though. I was part of a group that would be up to a lot of mischief. Especially with girls. Since our houses were small, the intimacy between our parents was not hidden from us. Whenever we got a chance, we would go and request girls to play house with us. Then we would hug the girls. They were also pretty happy to play this game. We would touch each other, pretending to be grown-ups.

Once we got into trouble because two or three of us seven-year-old boys lured a girl into a lonely wooded area and tried to get fresh with her. We told her a yarn about how 'ritha' seeds were growing out of the ground there. But that girl wasn't impressed. When she figured that we were up to no good, she complained to her parents. They in turn told our parents and we got a sound thrashing from them.

MY GREAT-GREAT-GRANDFATHER, Ganpat Ram Goenka, had three sons who became the founders of Mandi Adampur. One of them was Ram Gopal Goenka, my great-grandfather. Ram Gopalji had three sons—Jagan Nath, Gopi Ram and Inder Prasad. Of them, my biological grandfather was Gopi Ramji, whose six sons and two daughters included my father.

But Gopi Ramji's elder brother, Jagan Nath, did not have any children. As part of a filial arrangement, my father was adopted by Jagan Nathji. And therefore, I recognize Jagan Nathji as my grandfather or Dadaji since it was he who effectively brought up my father and the rest of the clan. He was the patriarch who influenced the lives of everyone.

He was my inspiration and guide as well. But I will come to that later. There is an interesting story that explains why my Dadaji didn't have children. Those days the local barbers were very active in matchmaking. One day a barber brought a *rishta* (marriage proposal) for my grandfather to my great-grandfather. But he refused, as he did not find it suitable. The barber felt affronted and taunted him. 'Will you bring a girl from the Chudiwala family?' he asked. The Chudiwala family commanded a higher status and was more affluent than our family. My great-grandfather was a small businessman compared to them. Angered by this taunt, my great-grandfather pledged that his eldest son would marry a girl from the Chudiwala family, and made an offer to that family. Taking advantage of the situation, the Chudiwala family proposed a girl who was sub-normal. This girl would be my grandmother. She was a simpleton, a loving and caring person but with underdeveloped faculties. She would talk to herself and had strong views on everything. She saw everybody in sharp black-and-white terms.

She once had an argument with my great-aunt and, without realizing the consequences, hit her with a metal dish. My grandfather's sister died as a result of the blow. My grandfather could not accept the intellectually challenged person that was his wife. He never consummated his marriage with her. And also chose never to marry again. But he was keen on a child and adopted my father, who was the eldest of the six sons of his younger brother.

This was perhaps fortuitous for me as it changed the course of my life. For, my father and his brothers did not have the business acumen of their elders. They had intermittent success but could not build a sustainable business.

My father's brothers had left Adampur to set up their business in other regions but were not doing well. My father wished to mentor them. My grandfather told him: 'I have done a lot for your brothers. But their destiny is weak and they have made many mistakes; they have lied to their associates, they have cheated people. Are you God that you think you can help them?'

But my father stuck to his position. My grandfather took a tough stand: 'If you want to go and help them, I won't stop you. But I will not allow my grandchildren and my daughter-in-law to suffer with you. They will stay with me in Hisar. And I will take care of them.'

My father had three sons, including me, and two daughters when he decided to leave us. My two brothers, Laxmi and Jawahar, were two and four years younger than me respectively.

This was a defining moment in our lives. It was 1958 and I was barely eight years old. My father stayed away for the next thirteen to fourteen years. He would visit us but wasn't around when we were growing up. He stayed at Korba in Madhya Pradesh for many years. The National Thermal Power Corporation had a generation plant in Korba and the region also had a lot of mining activity. My father traded in copper and other minerals; he also became a transporter.

When my father left, Dadaji decided that it was time for us to move from Adampur to the neighbouring large town of Hisar. And so I grew up in Hisar, with Dadaji as the father figure in my life.

DADAJI WAS MY friend, philosopher and guide. A successful agriculture commodity dealer, banker and commission agent, he had a great influence on my upbringing.

The family had set up a few mills to process grains and lentils. Dadaji traded in grains on a commission basis. He would finance the agro-produce and was also influential among the traditional moneylenders. They would buy particular grains from the mandi where such grain was surplus and sell wherever the demand was. They would trade in different types of grains in different states.

Dadaji traded on behalf of others, too. Some of his clients were into speculative trading. They would place orders with our firm to buy 200 sacks or quintals of a particular grain and hold it for them.

There was no organized or legal commodity or futures exchange like we have today. We had to finance the clients, buy the grain, physically store it and then sell it. All the transactions were reported to the client on postcards and were recorded in the account books in the form of debit and credit.

When my father left, there were four key family members and personalities that were running the business. My grandfather, his youngest brother Inder Prasad, their nephew Ghisa Ram (sister's son) and chief accountant or munimji, Lakhi Ram.

Ghisa Ram was a scary figure for us three brothers. We thought he did not do much for Dadaji but had an undue influence on him. Ghisa Ram would help with matters of taxation and banking, as he was better educated. He got involved in local politics and was even elected as a municipal councillor. Even though Munimji was not part of the family, he was treated like one. For instance, the groceries for all four families would be bought together. His kids would study with us.

Our school would begin at 7 a.m. and end at 1 p.m. After school I would go to my grandfather's office. He would be sitting on a large white cushion or *gaddi*. The person occupying the *gaddi* was the owner and in charge. Others would sit around him and support him.

I worked essentially like Dadaji's assistant. He would ask me to make calls to his associates or clients. I would dial the exchange, and ask the operator for the number. He would connect and I would hand over the phone to Dadaji.

Then around 3-4 p.m. he would start dictating letters to me. I would write these down on postcards. These were the trading rates of various commodities like cotton and dals. It was like a commodity rate ticker service. Between both of us, we would write about forty to fifty such letters every day.

Amazingly, he would not keep a copy of the rates. These were memorized by him, his brother and by Munimji. Sometimes, at the end of the day, I would be asked to check the accounts. This would involve matching the day's trading activity with the cash at hand. We would calculate the money spent and earned. And then ensure that it matched the cash in hand.

All transactions would be scribbled on a sheet clipped to a hand-held board. One column listed cash that went out, the other column the cash that came in. The net cash would be matched with the money that the day had started with.

In the evening I would help in lighting small earthen lamps or *diyas* in front of the deity in the shop and home. My day would end around 5-6 p.m. And then I would happily run off to play. Games like *gilli danda*, marbles and hide-and-seek were our favourites.

My role as assistant to Dadaji was almost a given. I was the eldest grandson, so it was natural that I would be supporting his work at the *gaddi*. I was never ordered to be there. I think I enjoyed learning. It was exciting at a level to see my Dadaji at work.

My cousins were also expected to be at the *gaddi*, but no one spent quality time there. Even their parents did not encourage them to be at the shop. At times, I think this attitude cost them dearly in life. They could not match my success in business. Some ended up working for me or took up other jobs.

I MANAGED TO do well in school thanks to the grounding in math that my father gave me. Even though he was not living with us, he would teach me math during his visits. When we moved from Adampur to Hisar, I joined Jain Primary School in the third standard. The secondary education began from the fifth standard, when I joined Chandulal Anglo Vedic (CAV) High School, managed by Swami Dayanand Anglo Vedic (DAV) Education Institution. DAV had many schools and colleges under its management in north India. It was a large education society run and managed by the Arya Samaj community.

Just before I was to start class 5, my father was home on a visit. He told me that English would be taught in class 5, and asked if I would be interested in learning it from him. This was an emotional moment for me. I used to miss him. When he offered to teach me to read and write, I was delighted. My father did not see this emotionally but practically. Over a week, he taught me the alphabet and how each letter was pronounced. He tested my learning and was pleased that I had learnt what he taught me.

During another trip my father taught me how to add, subtract, divide and multiply numbers. He taught me the basics of mathematics in a way that I would never forget.

These two skills of being good in English and math gave me an advantage over my classmates, and even students who were one or two years senior to me at school. I performed well till I completed my high school education. I was usually among the top three students.

As I look back, I realize the importance of the smallest of efforts by a parent. Even a few days' support can put the child ahead of others in the school and the community. I believe that those lessons in English and math gave me such confidence in myself that it has kept me ahead of my peers till today.

Those days the kids of trading families also had to learn accountancy. This was not taught in schools, and special tutors would be appointed. This was called *munimi*. I did not have to take these classes, but my father did give me some smart tips in accounting. He taught me how to manage numbers and accounting with clever tricks. These lessons have always been handy in my business life, helping me calculate on my feet while conducting negotiations. I also learnt to write in Gurmukhi. Thus I could speak and write in three languages: Hindi, English and Gurmukhi.

While I was in school, my father did not send any money home, as his own business was floundering. Sometimes we felt a bit neglected. Though, on the face of it, the children of all three families were getting equal money per month of about Rs 10-15, our cousins would get more privately from their parents. My younger brother, Laxmi, was rebellious. He resented other kids getting more than what we did from Dadaji. At times, we sorely felt the absence of our father .

Our family grew since another son and later, a daughter, were born to our parents. We had a visiting father so my mother brought us up almost single-handedly.

I WAS ABOUT ten years old when I met a member of the Rashtriya Swayamsevak Sangh (RSS) in Hisar. He was not a *pracharak*, but was active in the Sangh. He asked my cousins and me to attend the *shakha* or group meeting that would begin at 5 a.m.

We were not sure why we had to wake up early in winters to attend the *shakha* before school. I knew that my father had been a *swayamsevak*. He used to run a *shakha* in Adampur, where I was born. Despite the irritation of having to wake up early, we started attending the *shakha* out of curiosity. I attended it regularly from classes 6 to 10.

Kids like me used to enjoy the sessions. The *pracharaks* would narrate tales of patriotism and mythology. For us kids it was like a session of storytelling. It would take us into a world that was fascinating and exciting.

Once I attended an RSS training camp under tense circumstances. It was a fifteen-day camp and was to be held some distance away from Hisar. Dadaji was on tour, my father was also away. I sought permission from my mother. She agreed. But then there was the problem of money for travel and stay. I asked Dadaji's brother Inder Prasadji for some money. He was in charge of the *gaddi* while Dadaji was away. But he refused. 'All this is useless. It is nothing. This will spoil you,' he chided me.

I was disappointed. But his daughter came to my rescue. She supported me and told me to ignore her father's anger. Some other friends offered to pay for my travel and stay. It cost only about Rs 30 to 50. I attended the camp and returned happily. But when I met Inder Prasadji and touched his feet, he stepped back. It was a rebuke to me. He was deeply upset about my defiance. Now I think even Dadaji would not have given me permission to attend the *shakha*. Dadaji was not fond of the RSS. He saw it with some suspicion.

Dadaji was not just a patriarch, he was a leading member of the community. He was the arbitrator of local disputes between traders, and fond of matchmaking. He led a business that included eight other brothers and their families. I would say that more than a hundred family members were dependent on him for financial, emotional or moral support. My grandfather's sister's husband had died at an early age. Dadaji also brought up the four children of his sister. He made sure that they received a good education. One of the sons became a civil engineer. It was a big achievement for the family. In the community Dadaji was called Lalaji. Others called him Bade bhai, Tauji or Dadaji.

Around bedtime, he would sometimes summon me to press or massage his body. Then we would chat for a while before he slept. He would tell me stories about the people he was dealing with. He would talk to me about relationships, behaviour, attitudes of people. I think I learnt a lot about people and their nature from these chats.

I graduated from working at the *gaddi*, and was assigned to the dal and cotton mills. I would count the material and enter the weight of each bag of raw material coming in and finished goods going out. The finished goods, like cotton, gram dal, and by-products such as cattle feed, were sold in different places like Delhi, Uttar Pradesh, and Punjab. The staff member experienced in buying and selling would go with the truckloads of materials to these places, and sell them through prefixed commission agents. I would accompany the staff member, and bring money in cash back to Hisar. This was generally a two-to-five-day trip.

WITH ITS CAPITAL at Agroha, the kingdom of Hisar was once a part of the Mauryan, Kushan and Gupta empires. During the Tughlaq, Mughal and British empires, Agroha was prominent not just because it was the capital but because it was the birthplace of our (Vaishya) community. Traders from the region were referred to as Agroha-waley. Over time this description was modified to Agarwal.

Thus when my forefathers came to Mandi Adampur in Agroha region, they were actually returning to the land from where our community had originated. I feel proud and fortunate to have been born and brought up near Agroha.

As the business grew in Hisar, I was given major responsibilities by Dadaji at a relatively young age. He had much faith in me and was keen that I get practical experience of the trade. I was twelve or thirteen years old when I was asked to accompany trucks carrying our goods for sale to Delhi and other destinations. I would arrive in these trucks at the offices of the commission agents in these markets. The agent would sell the goods. I would oversee the sales transaction; collect and count the money, and return by bus to Hisar.

It was unusual for a young teenager to be given such work. Dadaji must have been aware of the risk. I don't remember the exact reason, but the first time he asked me to go was because no other suitable person was available. But since I handled the work without any problem, Dadaji felt confident about future trips.

Sometimes, I would stay at the sales destination for two to three days, alone. Usually, I would carry Rs 5,000 to Rs 15,000 with me on my return. My family had confidence in me and I wasn't scared either. But looking back, I realize how risky it was for a pre-teen boy to be travelling with such amounts of cash. In many ways, it was part of my preparation for responsibilities ahead.

Delhi is 160 kilometres from Hisar and an important market. I would accompany the employees or munims to Delhi to sell grains and other commodities. A private bus service called Krishna would start at 5 a.m. from Hisar while the return journey began at 5 p.m. from Fatehpuri in Delhi.

The journey was four-and-a-half-to-five hours with a single break at Meham, where the driver and passengers would grab a bite. I would enjoy my meals on the highway. For me each trip was a unique experience, I would learn something new on each journey. For a small-town boy, travelling to a big city like Delhi was a delight. I would share my experiences with other kids when I returned. Apart from Delhi I would travel to Ludhiana, Hoshiarpur and Pathankot. We would go to mandis to buy and sell commodities.

The most important task of the day when we were on the road was executed in the evening. Each person had to tally their expenses before going to bed. Every single paisa had to be accounted for. The tiniest expenditure, like rickshaw hire cost, would be put down on paper. So if I left Hisar with Rs 100, and at the end of the day I had only Rs 57, I had to know where I spent every rupee and every paisa. We could spend on entertainment, such as movies, as long as we were honest and wrote down the cost in our expense list.

I remember an interesting example of how flexibility in these rules was offered to loyal munims who had to remain away from their families for months. Once, when a munimji returned, I was asked to take a report on the expenditure from him and submit it to my grandfather. This particular munimji was very meticulous. He started dictating the list of expenses for about two months of his travel. I was dutifully noting these down and tallying the figures.

One of the items he listed was Rs 15 for 'change of oil'. This was a bit perplexing for me, as he did not use any vehicle.

So I asked him, '*Munimji, ye kiska tel badalwaya tha? Koi truck tha kya?*' (What was the oil change for? Was it for some truck?)

He looked away and did not reply. But my grandfather heard.

He gently but firmly told me to write it down but not question the munimji. The matter ended there for the moment.

But the question remained with me. I couldn't understand why a meticulous person like my grandfather would allow a suspicious-looking expense without questioning.

The answer came to me some days later. The munimji had been away from his home and family for more than two months. So this sum of Rs 15 had been spent as entertainment. But this was not the cost of some movie or mela. He had probably visited a kothewali, a lady of the night. Such expenses were allowed as Dadaji was pragmatic about it. The only issue was that such expenses had to be listed under other categories.

To earn some money on the side, I started teaching students of classes 6 to 7 when I was in class 9. I earned about Rs 15 per month. But it did not last. As soon as my grandfather heard of it, he asked me to stop. Although we did not have dearth of money, my grandfather controlled the income of the family. He gave Rs 150 to my mother every month for all expenses. My mother would not talk to my grandfather directly, according to prevailing customs. She would ask us to inform him whenever she ran short of money. Dadaji would direct the cashier to give more money but he would chide us at times and caution us about needless spending.

Money for monthly expenses came from the general business kitty managed by Dadaji. This was not his personal money but of the *gaddi*. In our system the *gaddi* was the seat of power, a symbol of the business.

Dadaji being strict about spending, even simple entertainment like watching a movie was not allowed. Adampur did not even have a cinema hall. I must have watched my first movie in Hisar. We could watch movies only in special circumstances. If a son-in-law of any of our uncles would visit, we could accompany him to the cinema. We could go as escorts of visiting guests or relatives.

I think my first movie was *Aaye Din Bahaar Ke*. Rajshree was my favourite heroine. I enjoyed the movies but did not like the restrictions on watching them. A relative and I devised a clever way to watch movies. If we were away from the shop or home for three hours, there were questions to be answered. How could we stay away and still watch a movie? I collaborated with Satya Narain. He was my age, even though he was my father's cousin. We would buy one movie ticket and watch one half at a time. I would watch the first half and return home after one-and-a-half hours. He would watch the second half. The next day we would swap the roles and watch the other halves. This way we would not be away long enough to attract suspicion. There were only two cinema halls and both were less than ten minutes' cycle-ride away. We could return to work quickly since the halls were so close.

Once in a while we would get caught and had to face the anger of Inder Prasadji. Dadaji would support me. '*Chhottey*,' he would tell his brother. '*Chhora kaam karey hai. Kya baat ho gayi gar picture dekh li.*' (He is a hard-working boy, so what if he went to see a movie.)

IN 1965, AT age fifteen, I passed the tenth standard with good results. Since I had been good in math and accounting, I was keen on doing further studies. My exposure to other cities had also created an urge to do something new. I told Dadaji that I was keen to do a professional engineering course.

He encouraged me and helped me apply to a college in neighbouring Sirsa. I was admitted in a professional overseer's course at Punjab Polytechnic Institute. It was a three-year diploma course of engineering that prepared students for higher education.

It also allowed the graduate to become an overseer/junior engineer in the government. This position was one level below engineer in the overall structure of bureaucracy. After overseer, the next level of officer was called a sub-divisional officer or SDO in local government. This was an important position. It promised an important career with power and influence. Very few students from the trading community took up professional courses. Most joined their family businesses.

My admission into the course was considered a rare achievement. Especially because my father had not studied beyond class 10. Most girls in the family did not study beyond class 8 or 9. Dadaji had studied only till class 4 but was fluent in writing and speaking in Urdu.

Dadaji was not keen to see me go. Most patriarchs prefer to

see their sons stay in the same business. Unlike others, though, he had a broader view of life and did not stop me from seeking a new profession. He even wrote a letter to a business associate in Sirsa to look after me. My father wrote an encouraging letter to me also. He was happy that I wanted to become an engineer.

But my granduncle Inder Prasadji was not happy. He felt that I would have been more useful if I had joined the business. Instead, I was taking a course that would be useless to the family trade. This type of thinking had stopped many children of our community from seeking new careers.

Nevertheless, I went ahead with my plan since my father and Dadaji were supporting my decision. I took a room on rent and shared it with a schoolmate, Radhey Shyam, as I did not get a place in the hostel.

My earliest friend at the polytechnic was Ram Singh, known as a tough guy. He started calling me Lalaji since I was from a trader family. I would draw about Rs 300 per month for my expenses from my local guardian. But other students would get only about Rs 100-150. For them, I was a rich kid. This amused me since I never considered myself affluent.

Poor or rich, the source of entertainment and interest for all students lay outside our polytechnic. The National College near our polytechnic taught arts subjects and was co-educational. Legend was the girls at this college were very beautiful and courageous. The reality was that very few colleges had girls those days. So girls in any college were considered a great combination of beauty and brains. For all the boys at the polytechnic, National College was the place to see and meet 'modern' girls. To hunt and track down girls, most of us bought bicycles. The first thing we would find out about a girl was her residential address. But once we got to know that, we had little clue about the next steps required to talk to a girl, let alone ask her out.

Soon I started hanging out with a bunch of students led by Ram Singh. This bunch enjoyed their parties and frequently got drunk. They also loved meat dishes, which I avoided being a vegetarian. My roommate and classmate, Radhey Shyam, did not approve of these habits and became my conscience-keeper. He was from a family with modest means and did not have money to waste on such indulgences. He would keep warning me about drinking and ensured that I did not fall prey to the temptations offered by Ram Singh's wild ways. Even though I hung around with such friends, I refused to drink or eat meat.

Radhey Shyam also taught me to cook and survive on my own. I would make roti and sabzi. Before that, I had never entered the kitchen at home. Even though I learnt to cook, I did not actually start cooking every day. I ate at a local dhaba since it was cheaper. I found cooking to be a waste of time.

I made a lot of friends in Sirsa. In some ways it was a coming of age for me. Living alone and interacting with strangers helped me find my voice. I learnt how to converse and engage people. Professors would ensure that all students spoke in class. This helped me become confident about speaking out my mind in a group. Academically, however, my first year was not very good. I had to reappear for exams in two subjects.

It was also a period of sartorial transformation. From shorts and shirts to trousers and full-sleeved shirts. I even had my first suit made. I washed and ironed my own clothes. And enjoyed the freedom of shopping for my own wardrobe. I did not follow the ruling fashion trends set by movie stars.

I went to Bhatinda a couple of times with my friend Ram Singh to see a travelling circus. I was excited to see the acrobats and the wild animals. But what really excited us was a sight we had never seen before: girls in bikinis and tight clothes.

2

A CRISIS IN THE FAMILY

I start work as a teen to pay off debt

THE GOOD TIMES at Sirsa did not last. I had to return mid-way through the second year. While I was away, our family business had suffered great losses. I didn't know then that this would be the start of a long, traumatic period for me and my family.

Dadaji wrote me a letter asking me to return, as he could no longer afford the expenses of my stay and studies. It was a terrible feeling. I cried when I got the letter. Dadaji was our icon and could do no wrong. How could someone like him fail in business? It was beyond my comprehension.

When I returned, I had another shock in store. Dadaji was the not the same person I had left a year ago. Earlier, he would be unfazed by business losses. While my uncles and cousins would be terrified of downturns in the business cycle, Dadaji would stay unperturbed. I never saw him worried or nervous. He had a strong, commanding voice.

'*Gir pad key hi seekhtey hain. Nahin to ghursawar kya hua,*' he would say. A horseman learns to ride only after he has fallen a few times. That was his credo, his confidence. He was not afraid of taking risks and was happy to take on extra responsibilities. He was not weighed down by the pressure of more than a hundred family members being dependent on him. He led a disciplined life and approached his business with great discipline and focus.

That's the Dadaji I had left in Hisar when I went to Sirsa.

One year later, he was a different person. I could barely recognize him. When I met him, he was quiet and introspective. The booming voice had been replaced by a soft pleading voice. He appeared to be someone who was scared of his own shadow.

He was fifty-four or fifty-five years old when his business began to crumble. After decades of success, the person who was our patriarch had become a shadow of his former self.

There were two reasons for our losses: the setback we faced in the *satta* bazaar and a high degree of fluctuation in cotton prices.

Satta bazaar was a forward trading market for grains and commodities. It was not an illegal activity and it was common for established traders to participate in the *satta* market. As in any forward market, traders hedged their bets for future market movement. Unfortunately, some of the trades that Dadaji made in cotton were speculative. The price movement went against his bet and he lost over Rs 50 lakhs.

The second big reason was the family's investment in an oil mill. We invested Rs 35 lakhs in the oil mill but it sank us. The oil mill required high working capital that we could not manage. We had taken a loan of Rs 20 lakhs, but our family had invested the rest of the money in it.

Every business came to a standstill. Oil extraction and dal processing could not be continued. Even trading activity came to a halt. Within weeks word spread that M/s Ram Gopal Inder Prasad (the name of the main firm) had failed to honour its commitments. For the market it meant that the firm had become bankrupt.

As a result, the working capital and the cash being rotated in our trading activity reduced sharply. We had to buy grains in advance and get the sales receipts later. If any trader did not have enough money in play, he would not be able to buy in advance and do business. Oil mills and dal processing units also needed working capital. The lack of cash flow led to closure of these units. Dadaji seemed unable to figure out a way to break this vicious cycle.

There is a proverb in our community, '*Occhi punji, khasam ko khaye.*' A cash crunch can kill a businessman. Our annual revenue those days was Rs 3-4 crores. We owned and managed two cotton ginning units, two dal mills, one oil mill and a commodity trading business that had branches in six towns. Though the turnover was Rs 3-4 crores, the profit margin was thin. Most years we would earn Rs 3-4 lakhs.

Our expenses were also substantial as the family was large. The marriages of many daughters of the family had been organized but some were yet to happen. The expense for each girl's wedding was Rs 30-40,000 and could go upto Rs 1 lakh. Our family was well known; we had a reputation to maintain. We did not cut back on wedding expenses.

Not only was the working capital gone, but we had debt. Dadaji did sell some assets to pay off some of it. He had a pink Morris, too, though I don't remember what became of it. He had to dip into his savings.

Our troubles only worsened. Dadaji's key business aide, Ghisa Ram, met with a terrible accident. He and my brother Laxmi were returning in a truck that was full of empty oil tins. They had sold the oil and were on their way back to Hisar from Delhi. The driver had been on the road for many hours non-stop and was sleepy. So Ghisa Ram asked him to take a tea break at a dhaba about 16 kilometres before Hisar. Ideally, Ghisa Ram should have allowed the driver some sleep.

When they resumed the journey, my brother Laxmi went to

the roof of the driver's cabin of the truck to lie down. Ghisa Ram sat in the front cabin with the driver. Unfortunately, the tea break had not helped the driver and he dozed off at the wheel. Before anyone could realize what was happening, the truck crashed into a stationary vehicle by the side of the highway.

Ghisa Ram was severely injured. He had multiple fractures in both his legs. Thankfully, there wasn't a scratch on the driver or Laxmi or any other person.

Dadaji depended a lot on Ghisa Ram. But Ghisa Ram could never fully recover from the injury. He had to be in the hospital for many months. This had a further demoralizing effect on Dadaji.

Dadaji always belived that hard-working people made their own luck. He was a *karmayogi*. A person who worked his way to success and who did not believe in luck. But the losses in business and Ghisa Ram's accident shattered his confidence.

He started buying lottery tickets, so that he could pay off some of his debts. Most people who visited him came to ask for monies they had lent the family. The pressure on Dadaji was immense. In desperation he started consulting astrologers. He showed even my horoscope to many of them. He went on a long journey to consult a famous astrologer and a Bhrigu Shastra expert. Watching him suffer was very difficult for me. I felt helpless and was desperate to do something to make Dadaji feel better.

One day I could not hold back any longer and spoke to him. '*Is tarah kab tak chalega, hame kam to shuru karna chahiye,*' I said haltingly. (How long can we continue or survive like this? We must start some work/business.)

He gave me a hard look and said slowly, '*Bina paise ke kaam kaise shuru karenge, tujhe to pata hai mandi mein koi udhar nahin dega, apna bazaar band hai.*' (How will you do any business without money? As you know, no one will give us credit for trade. No one will lend us cash for the business. Our credibility is very low.) He broke down as he said this. It was tough for him to appear weak in front of his grandson.

I was deeply hurt and shocked to see him cry. I said, *'Aap izazat den to ek dal mill to main chalana chahunga.'* (I would like to revive operations of one of our dal mills, if you permit me.)

He looked at me with disbelief on his face. How could a seventeen-year-old do business with no working capital. Especially at a time when there was a stigma of bankruptcy on our family? He broke down again. He was touched by my offer.

'No, don't start anything right now,' he said. 'These are bad times for us. Bad luck will not allow you to succeed.'

But I persisted. 'Please allow me, Dadaji. I promise you that if I can't make profits, I will ensure that we don't suffer more losses.'

I kept persisting till I wore him down. He wanted the tide to turn before resuming business. But I felt that unless we kept making an effort, the tide wouldn't turn.

After a while, Dadaji did not have the strength to say 'no' to me. When he agreed, he set me on a path of running my own business. It was 1967 and I was seventeen.

I HAD BEEN idle for six to eight months before I took Dadaji's permission to take charge of a dal mill.

During this period I went to visit my father's younger brother, who lived in Gangtok. He received me with great affection. He took me along to meet the king of Sikkim, called the Chogyal. On another occasion, he sent me to the border camps of the army. This was my first visit to the Himalayan mountains. I was fascinated. We partied at the army camps too. We would eat the food in the mess and drink from noon onwards for three to four hours. This was my first experience of alcohol. I don't think I handled it well. Sometimes I would sleep from afternoon till the next morning. I would wake up with a severe headache.

After this trip, I could not take such a carefree holiday for years.

I now set about reviving our mill. I not only had to do so without any working capital but also had to be responsible for the basic expenses of three families that had forty members. Even after Dadaji had sold some assets, we were left with a debt of Rs 5 lakhs. This was more than the profit we made over a two-year period.

A single dal mill could earn only Rs 50,000 to Rs 1 lakh per year, which was not enough to meet the expenses of the family. It actually left us with a small deficit.

I did not have much idea about running a mill, and realized I would need help. The one person I could think of for help was Chander Bhan. He had been a broker in the trade for our family for over two decades, from the early 1940s. He met me but was very sceptical about my plan.

The obvious first step was to procure raw gram or lentils. For this, all mill owners had to go to the grain market and buy the crop of gram brought by farmers. But since we did not have the money, we would have to buy the stock on credit. And since our business was down and there was no cash flow, no one willingly gave us raw materials and grains on credit.

Chander Bhan pointed out that when we bid for the grain, the shop-owner would not sell to us at our highest bid but privately sell to a credible buyer at a price lower than ours. 'This will be insulting,' Chander Bhan said, 'hence I won't like to accompany you.' But I was determined and demanded that he come with me. I was hoping that he would be able to convince the traders to give me some supplies on credit.

Chander Bhan came with me but things didn't go as planned.

He refused to speak to any of the dealers because he lacked conviction in my ability to pull it off. I was left alone since my only hope did not speak up for me.

To understand the challenge of starting the dal mill, one must understand the mandi system. In any mandi, the farmers bring their produce to the commission agent for selling. They all have their predetermined commission agents. The state governments have marketing committees (under the marketing board of agro commodities at state level) in each mandi. The mandate of this board and committees was to ensure that the farmers get to determine the true potential value of their product and decide of their free will whether to sell or not. The value per 100 kilograms of grain was arrived at through a bidding process by all potential buyers, and conducted in the presence of a marketing inspector, who was a government employee.

The highest bidder was chosen or rejected (in case the farmer did not want to sell at the highest bid price). After that the produce was weighed and the delivery was given to the buyer from the place of the commission agent. Each bid was 10 paisa higher than the previous one. If the farmer did not find the right price and chose not to sell, the grain would be repacked and stored at the commission agent's warehouse. At times, the farmer would wait for weeks and months for the right price.

Upset with the Chander Bhan episode, I decided to brave it out alone. I started attending grain auctions in the hope of spotting opportunities and convincing someone to give me the grains on credit.

But I was made to feel unwelcome and unwanted by the agents. They saw me as a member of a discredited family that was in debt. There were occasions when the agent would not allow me to buy even when my bid was the highest. The price of black gram was about Rs 30-35 per quintal (100 kilograms). Even though my bid was the highest, the commission agent said that the farmer did not want to sell. But the moment I turned away, the agent sold the chana (black gram) to the second highest bidder.

I tolerated this humiliation for a few months and began to feel suffocated in the mandi. I would attend the market everyday and observe the proceedings. The traders tolerated me but no one encouraged me to do business.

My desperation was growing. Even though all the traders and agents were my father's age or older, I decided to be aggressive with them.

Once I spoke loudly to an agent. *'Tauji, aap mujhe kyon nahi maal dete hain. Agar aapko paise doobne ka dar hai to maal delivery hone ke pehle meri gaddi se paise le aayey.'* I was loud so that other bidders as well as the marketing inspector would hear my plaint. I had offered one agent that he collect money from my office even before the delivery of the goods. This was to build confidence in them in my ability to pay for the purchase.

'Nahi, nahi beta Subhash, yeh baat nahi hai, main to paisewala dukandar nahin hoon, kisan ko to aaj hi paise dene padenge, aur tum agar 15-20 din paise nahin do to main to business nahin kar paunga,' the agent said. He could not wait for me to pay him in twenty days' time since the farmer had to be paid immediately. I had to reassure him that I would pay as per the existing trade practice, where the due date for payment was five days. This assurance started to work with other agents, too. But the reluctance remained for a while.

In these small mandis people know and talk about every small issue. I had become a talking point because I represented a financially weak business family, which was once very credible.

Also I was the youngest person in the business in Hisar mandi. Hence there was curiosity about whether I would succeed or fail in my attempts to restart the business. Most thought I would fail since the conditions were very tough.

I received unexpected but welcome advice from another broker who also happened to be a distant relative. Hari Ram Khariawala came to me uninvited and said, *'Tujhe ek salah deta hoon. Jab bhi tere paas dal bech kar paise aayen to tu commission agents ki dukano par ja kar paise de aaya kar. Jiske Rs 2,000 dene hon usey Rs 1,000 de diya kar or jiske Rs 5,000 dene hon use Rs 2,000-2,500 de aya kar. Aisa karne se tere par bharosa jamega.'* Pay everyone a little bit of what was owed to them as soon as there was cash in my hand. An immediate part payment works better than a delayed full payment. I would have to pay the entire amount, but would have to do it in phases.

I started to follow this advice religiously and it worked like magic. Slowly but surely, this improved my creditworthiness in the mandi. The commission agents started welcoming me and even grew fond of me. They would say, 'Subhash *to achha beta hai, iske paise ki koi dar nahi hai.'* They were now more confident of my ability to pay.

They slowly stopped coming to my shop for collecting their dues, as they believed that I would come to their shop to pay them as soon as I had sales realization.

My biological grandfather also came to my help. He had returned to Mandi Adampur where his creditworthiness was intact. He could buy chana from Adampur and send it to my dal mill in Hisar. This was very helpful, as I could take longer credit period from Adampur suppliers.

My father's brother-in-law gave me gram on credit and allowed us a longer credit period. He had started his commission agent's business in Hansi (a market close to Hisar).

My effort to convince the market about my creditworthiness paid off. It took a lot of patience, some luck and some proactive steps. I had to manage the cash flow very smartly. Everything I earned had to be redistributed almost immediately in a way that all my suppliers would be happy. Also, I had to use some of the funds to manage our household expenses.

The regular supply brought the dal mill back into profits for us and helped revive the credibility of the family. However, the income from this was just enough to meet expenses; reduction of old debt was still not in sight.

AN ADDITIONAL PRESSURE was building up on me. Some cousins who were my age had been married and now the family expected me to follow suit.

At this stage I could barely contemplate marriage. The dal mill had barely turned around. We still had a mountain of debt to pay. I wrote a letter to my father and said that I would not marry until the family debt was paid, besides pleading with Dadaji not to force a marriage on me at that stage of my life.

I continued to explore new ways of earning money from any business venture I could think of. Soon, another idea struck me and I decided to try it out.

While our dal mill was operational, another unit that processed cotton and crushed seeds for oil was lying idle. I planned to use that unit for earning something extra.

The oil mill was called Adarsh Cotton Ginning and Oil Industries. It had a steam boiler for the process, which used coal as fuel. All such units in Hisar were allocated certain quantities of coal to run their mills, as there was a severe shortage of fuel.

The allocation of the coal was done by the Cotton Ginning Mill Association based in Bhatinda. I travelled to Bhatinda and met the association officials and asked for our share of coal.

It was an audacious move on my part to try and get coal for a unit that was not functional. It was a bit unfair to seek coal when I did not have a running plant to use it. But I was desperate. And I had a plan for using the coal.

The association doubted my demand but could not deny me the coal since every registered member had a right to ask.

To my great delight, I was allotted five wagons of coal for our mill. Coal was in short supply and I did not have any use for it. This is when my plan kicked in. I took the coal and then I traded it further with a higher margin of profit.

On each wagon I made a profit of Rs 3,000. As a result of this simple move, we got a neat kitty of Rs 15,000. This was a small but clever move to earn money.

I thought of several such small creative schemes to boost our profits. For instance, we rented out our truck to other traders. Each trip on the truck would earn us a profit of Rs 200 to 500. Such moves kept bringing in small quantities of cash to keep our hearth warm.

I started buying jute bags by the truckload from Delhi in another venture to earn profit to fund the mill. I sold these jute bags to dal mills and other traders who needed gunny bags for packaging their purchases of finished dal or other grains. One of my distant maternal uncles, Daulat Ram, got me longer credit periods from jute bag traders in Delhi.

My younger brothers, Laxmi and Jawahar, were always with me, helping at the mill or in the mandi. We worked as a team and divided all the work amongst us so that we would not have to hire anyone else. We would even lift heavy sacks and do other menial work to save the cost of hiring a labourer. By picking up our own load we saved labour cost of Rs 10 per day.

Accounting and management was done between us, too. Jawahar, who was good at math and money management, started handling our accounts. Laxmi was physically strong. He would handle the heavy work and manage the workers. He was in charge of the production and conversion process, where whole gram was dehusked and split to be ready for consumption in kitchens.

By working together and profiting from the work we did, I gained confidence about my own abilities to manage a business.

I had become obsessive about paying the debt, and ensured a strict repayment schedule. As a consequence, I felt strong enough to borrow some more.

I could also manage my relationships with friends and acquaintances. They would trust me with their money. I would borrow from these friends, who owned small businesses. I would borrow when they had extra cash and return whenever they needed it.

I remember many of them as they helped create a complicated but critical cash-flow process with me.

The irony was that the people who helped me with working capital were not my family or relatives. They were all friends and acquaintances who had faith in me. A few members of my family let me down. My mother's brother, Ram Kishan Khariawala, refused to help even though he had the resources. My father, though, had returned to support our efforts.

Within a year of restarting the dal mill, we managed to save Rs 2-3 lakhs. It was a decent profit but still not enough for us to repay the debt and have enough left over as working capital. I was about eighteen and amongst the youngest traders and mill-owners. With the help of my brothers, I had brought our family back on track, though the debt remained. We were also happy that our father had come back.

3

THINKING OUT OF THE BOX
Convincing FCI to adopt a new process

I HAD DONE everything I could to revive the mills and repay the debts. I was becoming a successful dal mill owner. There were many trader families who spent their lives as mill owners. I could have continued in the same way and perhaps started a second mill but there was limited scope for such business.

A chance meeting with an officer from Food Corporation of India (FCI) changed the trajectory of my life. It set me on a path that would take me to Delhi and help me climb the ladder of business success.

I met Chaudhary Aman Singh, an assistant manager with FCI, on a bus ride from Hisar to Delhi. He was much older than me—in his late forties—but we started chatting.

By now I had a good understanding of the grain trade in the region. The grain market and its traders were small players. Hundreds of thousands of such players were located across the region. The giant was FCI. As the official procurement agency of the government, it was the biggest buyer and seller of grains. Its decisions to buy, sell and price different grains would totally shake up mandis across the country. Though Aman Singh was an assistant manager, he was important enough for me. For some months now, I had been thinking about the way FCI operated. I had some ideas about opportunities that existed but had gone unnoticed. I kept inquiring about the procedures at FCI and used the information to develop the concepts I had. I took a chance and decided to share my ideas with Aman Singh.

'FCI buys various dals and foodgrains from the market under the farmers' support price programme of the Government of India. Then it stores for a couple of years and finally auctions it to traders who supply it in the open market. On the other hand, the Indian Army buys grains of various types through the ministry of food and agriculture. They float tenders and the traders participate and supply to the army. FCI is also part of the same ministry of food and agriculture, so why should FCI not supply to the Indian Army directly?' I posed this question to Singh. I had been thinking about this for weeks and was desperate to push FCI on this.

Basically, FCI was procuring and storing grains. It would again sell to the wholesalers for processing and selling the final product in the open market. Singh was an experienced official who knew the reason for FCI not supplying directly to the Indian Army. 'You see, our grain in the warehouse is always in raw condition and supply to the army is processed or upgraded grain, we (FCI) cannot meet the specification of the army, which is very strict,' he told me.

This is what I was waiting to hear. Without thinking twice, I made a suggestion to him that required both confidence and bravado.

'We will process all of these grains to army specifications against payment of processing/milling charges,' I offered.

He was surprised by my suggestion. While he liked the idea, he wondered if a teenager trader like me could actually deliver on such a proposal. However, he seemed impressed by my confidence.

'This is a good idea, but I am a small official in the entire process. Let me take you to some of my friends who work in our zonal office as well as head office in Delhi. They can arrange for you to meet some senior officials.'

My suggestion was simple. We would take the grains from the the FCI warehouse, process it and then supply to the army from our zonal area. All we would charge was the processing and labour charges.

This removed two or three levels of intermediaries and therefore also reduced the cost of delivery for FCI and the army.

The model I suggested suited me, as I did not have the capital to buy the grains for processing. The grains would be bought and stored already by FCI. All I had to do was to cart it to my mill and then deliver the processed grains to the army depot at a pre-determined fixed charge.

It was not possible to meet the officials immediately but Aman Singh and I kept in touch and met frequently on his visits to Hisar/Delhi.

He had made a proposal based on my idea and sent it to his supervisors. He hoped that sooner or later, they would realize its importance and take a decision on it. While we waited for the proposal, he ended up playing another important role in my life. He taught me to drink alcohol. He would drink every evening and invited me to join him. Initially, I refused, but had to give in on a day when we heard about a very positive development on my suggestion.

The proposal had not only reached the highest office at FCI, it was about to be recommended to the ministry. The innovation suggested by me and supported by him was actually being seriously considered by the ministry of food, Government of India.

For the first time, I had a peg of whisky with water. Not surprisingly, the whisky had a strong effect on me. I felt that I was soaring and flying above the clouds. I returned home on my bike in great spirits.

The policy was finally approved but as a pilot project only. Singh and I had to work hard to convince other officers about its success. Even though it was approved as an experiment, we had to ensure that it was a success so that it became regular practice. I was also keen that the work should come to me since it was my idea.

Singh helped convince the FCI officers to give me the work. We had to give them some financial incentive to allot the pilot project to us. At age eighteen, I had managed to convince a government-owned giant like FCI to adopt a new process.

We began with a large quantity—about 1,500 tons—of gram to be processed and converted to dehusked, split dal. FCI would deliver the grains to me in batches. I would process them and earn my profit. FCI delivered the grains processed by me to the army. This experiment lasted a year and helped me earn good profits.

FCI found this process to be more efficient and decided to expand it to other commodities. It started calling bids for processing on a regular basis. Meanwhile, the original suppliers to the army were upset since they had been cut of the process. They finally had no choice but accept the new process. They harboured a grudge against me and tried to dislodge me from the auction.

As they had been working with FCI and the army for many years, they understood the system better than me. First, they tried to get the grain processed by me rejected for poor quality. But they were not successful. Then they succeeded in convincing the finance department of FCI to add sales tax to the cost of supplies made from states other than Delhi. Those days, Delhi had no sales tax on foodgrains and by adding tax on outsiders, our cost would go up since we were supplying from Hisar (Haryana).

For a while, I thought they had succeeded and managed to push me out of this trade with FCI. Then I realized that I would have to move to Delhi to reduce the tax burden. Here again, my life took a turn. I did not have a choice, but it was a big decision for me. I would have to leave my family, especially my mother, alone and run my trade from Delhi.

WITHOUT THINKING MUCH, I moved quickly to set up base in Delhi. At the next bid, I surprised my competitors by quoting Delhi as the place of supply. My competitors were aghast. They had not imagined that I had the capability to move out of Hisar and settle in Delhi. They thought I was a small-town boy who would give up once he faced unfair competition. In a way, they were right. I had no idea about Delhi but I decided to take my chances any way.

To fulfil the tender requirement I needed a mill in Delhi. I could not afford to buy a mill so I hunted around to see if one would be available on rent. With the help of some trader friends and contacts, I found and rented one dal-processing mill in Nangloi area and another one at Nazafgarh in Delhi.

These mills were at opposite ends of the city and it would take me almost a whole day to visit both of them.

I also rented a *gaddi* (trader's office) in Ramesh Market near Chawri Bazaar. To rent these properties I needed more capital. While I finalized the deals, I kept racking my brain as to how to raise this money.

I had been obstinate enough to promise supplies from Delhi and had even rented the mills. But I did all this without having enough money. I felt that if I went ahead, I would be able to find a solution. This was a big risk for me. If I could not raise capital, I would lose the factories, lose the bids and go back to square one in my credit reputation.

I had brought only about Rs 50,000 from Hisar from the savings from dal mill profits. But I had an advantage. I won the bid and there was some time before the supplies would begin. After trying several options I felt that the best would be to join someone who had the resources. I chose to go with someone who had been competing for the same business but had lost the bid.

I talked to a competitor, C. Lal, who was influential in Delhi and had money to invest. He was assisted by his working partners, B.D. Hansaria and B.R. Chopra. We agreed to become 50:50 partners to process the bid. He would invest the money and I would be responsible for the entire business. We would share the profits equally. Lal deputed an accountant to manage the finances. I was busy almost 15 hours a day seven days a week. I had to single-handedly manage all process of business as well as FCI, Delhi (which was new and different from its regional operations at Hisar). I had to manage the mills, the freight at railways, the quality control of FCI and so on. After I had completed two to three new contracts I discovered from the accountant that we had made a small loss instead of profits.

I was shocked and told Hansaria that the accountant had made a serious mistake. By my calculations, we should have made a profit of around Rs 3 lakhs. But Hansaria did not agree with me. As the accountant was his and not mine, I felt helpless in the situation. I had worked hard in Delhi for a year, but all that effort seemed wasted. In my desperation to service the bid that I had won, I had been careless in keeping control over the accounts by trusting my partner. Here my inexperience in dealing with a bigger business house hurt my confidence and my finances. I had not been able to protect my savings. My business was lost. My hard work had failed and someone had made a fool of me. My partner seemed to have tricked me out of a few lakhs' profit with the help of his accountant.

I was dejected and didn't know what to do next. My failure in Delhi led to tensions at home as well. My family had been proudly supporting my success in turning around the dal mill. They had been a bit sceptical about my move to Delhi, though. When I went to Hisar after my humiliating loss, they called for a meeting. My father and brothers met me on this issue. The bid in Delhi had been managed by me alone as my brothers had remained in Hisar.

I accepted my mistake and my error of judgment. My family was sympathetic but felt I was not ready to work in the big bad world of Delhi traders. They felt, rightly so, that I was out of my depth in Delhi.

They asked me to wind up and return to Hisar and continue working as before. Reluctantly, I returned to Hisar in 1969 after living in Delhi for a year.

I continued working at the mill but I could not stomach the thought of doing the same work for the rest of my life. After seeing the possibilities of working with big companies like FCI in Delhi, I was itching to leave Hisar. Even though Dadaji had ordered me to stay there, I could not bring myself to accept the decision.

After about six to eight months, I realized that Hisar would not allow me to earn enough to repay family debt or grow the business—the profit margins would remain small. I wanted to do big things in life. I was dreaming all the time. I had to go to the big city to swim in a larger pool, and expand my horizons even though the immediate goal at that time was to get rid of the debt. I knew the grain trade instinctively, but I felt confident enough to go beyond agri-business. Despite the failure I knew how to get things done. My failure was to be cheated by an accountant and his boss. But I had been successful in servicing the bid and meeting my commitments to FCI. As a first-timer, I had done well enough to do business in Delhi.

One day I told Dadaji and my father that I was returning to Delhi to try my luck again. This came as a surprise to all as none of my family members was in favour of my leaving home. I had wasted one year and Rs 50,000 in Delhi. My family tried their best to change my decision. There were marathon meetings in the house to dissuade me.

They wanted me to stay in Hisar, get married and settle down. 'There is enough income from rents, land crops and so on, which is sufficient to live happily,' they said. But I almost shouted, 'What about the debt we have on the family, how will we repay that?'

This process went on for a week but I had made up my mind. They were worried about more losses. So I said that I would not take anything from them.

Before leaving for Delhi, I spoke to Dadaji. 'I am frustrated, Dadaji. I can't live in Hisar as there is not much to do.'

Surprisingly, he was the only one to support me. I had expected him to question my decision. He agreed with it. 'There is no point in you staying here. But if you are leaving, then go alone. Don't take anyone from here with you.'

This was an important and perceptive piece of advice from him. He was referring to my father and brothers. He wanted me to strike out on my own and use my best judgment. He did not want my decisions to be affected by other viewpoints. This would also help me believe in my own abilities.

So in 1970, at age twenty, I left Hisar again with just Rs 17 in my pocket. I had mixed feelings. I was angry and hurt. But I was also determined to get out of Hisar. I could smell freedom and a chance to explore new possibilities. For me succeeding in Delhi was all that mattered. I didn't know what I would do, but when I left Hisar by the 5 a.m. Krishna bus service, I felt positive about my future.

4

YOUNG MAN ON THE MAKE

Living off bluff, bluster and gumption

WHEN I REACHED Delhi, I went to Mangat Ramji, a broker based in Naya Bazaar, who used to be our neighbour in Hisar. I asked him for support and shelter. He agreed to help me because I was from his town. This was the kind of relationship that people maintained with each other. Mangat Ramji allowed me to stay at his office (which turned into a sleeping place at night) and even took care of my meals.

My first instinct was to continue working with FCI. That's what I knew best. I could not imagine doing anything else at that time. But for being eligible for FCI work, I had to take care of some prerequisites.

I decided to register a proprietorship firm and applied for FCI's tender. I used Mangat Ram's help and his address for the firm. But even though I was operating alone, I named the company Subhash Chander Laxmi Narain. This was the original name of the family firm. I registered the same in Delhi. This was to include my family in the name and also to give the impression that it was not a one-man company. The paperwork for setting up the firm was not too difficult.

To be eligible to do foodgrain business of any kind, including processing the FCI grains, I had to be registered and had to become a licensee with the food and civil supplies department of the Delhi administration as a grain dealer.

I went to the Naya Bazaar traders' association and met the secretary for advice on registration. Thankfully, I hit it off with him since he was also young like me. He helped me become a member of the association and, in turn, apply for a licence. Soon, I was a registered dealer.

Around that time FCI placed a tender for processing of 2,000 tons of chana dal for supply to the army in the Haryana region. This was being done under the same process that I had helped initiate. After the pilot project, the new process had been adopted for all tenders.

This bid was being handled by the regional office of FCI at Chandigarh. Most of my competitors were established and affluent. When such tenders were announced, they would fly to the regional offices of FCI on the day of the bidding. Unlike the pilot project, I was facing a much tougher competition here. But I decided to bid for it. I thought that even if I was not successful I would learn how to manage bigger players.

I discovered some of the tricks of the trade deployed by the other contractors. One contractor, named Anil Chanana, usually won more than his share of contracts. He had formed something of a cartel with other bidders. They would bid at certain prices and whoever won would sub-divide and share the work with others.

For this bid I reached Chandigarh a day before the auction. I could not afford to stay in a hotel so I stayed in a government-run guest house called Panchayat Bhawan. This facility allowed people from villages to stay at basic costs in the city. I would pay just Rs 1.50 per night for a clean bed, bath and a cupboard to store my suitcase. Another Rs 1.50 was the cost of a single meal. To get this cheapest rate possible I used my residential address of Mandi Adampur and not Hisar, as the rate for people from smaller towns like Adampur was lower.

I bought the tender document a day before and then met the assistant manager of the commercial division to understand the process. This division handled the bidding process for commercial grains. Apart from staple grain of wheat and rice, all other grains were called commercial crops by FCI.

I chatted with the officials and then asked them to join me for dinner the same evening. I was free and didn't have much to do. Moreover, I wanted to get to know them better so that I could develop a stronger relationship with FCI.

Surprisingly, the assistant manager agreed. He was a Sikh gentleman called Tibana, and he invited his assistant, Gurdeep Singh, to join us. We first had drinks at his house and then went out for dinner. I paid for the drinks and dinner. The money I had saved on my accommodation had been well spent on the dinner with these officers.

With their help I could understand the process. I also learnt about the cartel being run by Chanana and his group of associates. Now I had a simple strategy to win the bid. Since all the traders would stick to a minimum price, I would quote a figure lower than that. I was not part of Chanana's cartel so I could bid lower. When the bids were announced, I won the tender, much to the shock and surprise of the established players. Now I had to start work on processing the 2,000 tons of dal.

Gurdeep and Tibana continued to give me advice on future bids too. Over time we became good friends. I must admit here that they gave me an unfair advantage on a few occasions. They adopted a modus operandi that would ensure that I won the bid. Both of them were key officials who would open the bids in the presence of all bidders.

While I could upset the cartel on the first occasion, it was not going to be easy for future bids. But Gurdeep and Tibana ensured that my bid was the lowest. I would submit the bid without writing any figure to be charged by my firm as processing charges. The price being quoted was the labour charges per ton. Each bid would be Rs 7 to Rs 10 processing charge per quintal or 100 kilograms.

My bid's sealed envelope would be opened last and Gurdeep would write a figure for the processing charges that were slightly lower than the lowest bid. He would then declare my bid as the winner. Though other bidders were surprised at my wins, they could never smell a rat. Sometimes, Gurdeep would even put a figure that was 30 paise or 40 paise below the other bids.

I did not bribe them, but they took a liking for me because of my youthful ambition. And I can admit that I did help them in cash and kind whenever they asked.

Most other bidders thought I was just very aggressive and that it would be tough to compete with me. These bids would rank as my most important early wins. There was no looking back after the first win of 2,000 tons.

That first bid was a big gamble for me—I had bid for the tender even though I did not have any processing facilities. I did not even have the funds to rent a mill. My entire focus was on winning the bid; I did not have a logical plan. If I won and could not process the grain, I could have been blacklisted by FCI. I knew this, but went ahead anyway.

The risk paid off for me since I did manage to get a processing mill after I won the bid. Here again, I had to bank on my friends and supporters from Hisar. I borrowed money from Mangat Ramji and some other friends from Hisar who were in Delhi, to put together about Rs 40,000. I was known in the community from the time I had previously spent in Delhi for FCI work. This association and my previous track record helped me raise funds, which I used to rent a processing mill in the Lawrence Road industrial area. To save money on my living expenses, I decided to live within the mill. The environment was not very pleasant, but I had no choice except to reduce my expenditures.

This mill in Delhi became my base. I hired labour to work on the processing. And started bidding for similar tenders of FCI across India. I won similar bids in Jaipur, Delhi, Bhopal, Lucknow, and even in Bombay. I had understood the way FCI worked and won through aggressive pricing. This business of processing for FCI worked very well and I managed to pay off about Rs 3-4 lakhs of our Rs 5-lakh debt. This was entirely out of my business and efforts. Here I deployed a smart way of reducing debt that I had learnt from Dadaji. He had suggested that if I managed to pay off the principal, the creditors would be satisfied. The option I gave to the creditors was simple. Take the principal amount and forego the interest. The second option was to wait even longer for entire dues. In almost all cases the creditors would accept the principal amount to cut their losses.

Some people still fought for more, though. The last of these was a relative, Radha Kishan Kharakediwala, and his son. I offered to return the principal sum of Rs 50,000 but they demanded the interest, too. The interest on this was about three times the principal amount. I told them that if I paid them interest, all other lenders would want it, too, and then I would not be able to pay anyone. But Kharakediwala's son insisted. So I did a deal with them. I gave them an additional amount of Rs 50,000 and said that they could use this sum to earn interest. The loan had been pending for almost a decade. I told them they could return this extra amount to me after earning enough to make up for their loss.

They accepted the idea and took the money. They never returned this Rs 50,000, but I still gained since my interest burden had been reduced. This was a bit like restructuring of loans done by banks today. Both the debtor and the creditor benefited. With this transaction all my debts were paid.

Dadaji's deep understanding of people's psychology and

money management helped me reduce the debt. Dadaji had retired by then. He would spend his time in pilgrimages. I would visit Hisar every week or ten days to meet my family and my mother.

AT THIS TIME an interesting event took place in my life, something that required me to be street smart and think quickly on my feet.

Between 1971 and 1973, we had rented a two-room apartment in Rajendra Nagar in west Delhi. I was sharing it with two other friends from Sirsa. One weekend I was there with my friend Ramjas.

We were at a loose end on Sunday when our landlord, Makhan Lalji, dropped in. Rumour was that he used to deal in smuggled gold.

That morning he was in a generous mood and made us a welcome offer: 'Why don't you take my car and go for a drive around Delhi? The fuel tank is full and you don't have to pay a rupee.' My friend Ramjas got very excited and said, 'Bhaiji, let's make the most of this. I have to collect some electric motors from Bulandshahr (a town in Uttar Pradesh about 180 kilometres from Delhi). Let's go there and enjoy the drive, too.'

I thought it was a good idea. It was only 9.30 a.m. and we had the whole day to kill. So we jumped into Makhan Lalji's precious Ambassador car and headed out. But as we approached the Delhi-UP border, the car suddenly stalled. We were in for a shock. The Ambassador had run out of petrol. The fuel tank was empty.

I had Rs 15 in my pocket, which bought us only about 4 litres of petrol.

Then I thought of a solution. Some transporters I knew from Hisar had their offices at the UP border. We were two kilometres from their place. I told Ramjas that we would drive there and borrow some money from them. This would buy us enough petrol. We would return the money later.

We reached there and were met warmly by my transporter friends. While we were chatting, Ramjas kept prodding me to ask them for money. But somehow, I could not muster up the courage to do so. Instead, I got up after a while and returned to the car with Ramjas. He was very upset and pleaded, 'Bhaisaheb, let's return and ask them for money. We have nothing. How far can we go with so little petrol?'

I told him to stay quiet and to let me think.

As we passed Ghaziabad, heading towards Bulandshahr, I saw some persons trying to flag down buses and trucks to hitch a ride. In a flash I realized the solution to our problem. I stopped the car near these persons and said, 'I can take you, but you will have to pay Rs 15 per person. But the petrol is running low, so if there is a problem, you may have to push.' Surprisingly, two or three persons agreed. A comfortable ride in a spacious car was far better than riding in an open truck or a tightly packed bus.

And so I worked as a taxiwallah that day. We continued on our way to Bulandshahr, buying petrol on the way with the money we had collected. We completed our work there and headed back to Delhi. Again we stopped at various places, picking up and dropping passengers and charging them for the ride.

By the time we returned home, we had earned about Rs 60 over and above the money spent in petrol. Then we celebrated by buying a quarter bottle of whisky and enjoying a hearty dinner at Moti Mahal restaurant in Darya Ganj. We came to know later that the landlord had forgotten how much petrol there was in the car.

WITH MY FCI work becoming better, my mother asked me to start thinking about marriage. I was stalling, but once the last debt transaction was over in 1973, I agreed, and my father started looking for a suitable girl for me.

I was married on 4 December 1973 in Amritsar. There were more friends than family members in my wedding party. Of the 150 people who were in the *baraat*, about ninety were my friends. I had invited everyone who had done business with me or helped me in any way. For instance, I made sure that the young secretary of the Naya Bazaar Association attended my wedding. Many officials from FCI were special guests in Amritsar. My in-laws thought we were a disciplined *baraat* and the party was considered to be very civilized. It was only after we had left Amritsar with the bride that they discovered the empty bottles of alcohol at the hotel. My friends had managed to enjoy their drinks but still remained disciplined enough to impress my in-laws.

While the wedding went off well, the real drama had occurred during the finalization of my marriage. As in many situations, I met people who would play a critical role in my life in very unusual situations.

Before my family had met my wife Sushila's family, they had received another proposal from Rohtak. Our family friend, Har Gopalji, had brought us this proposal. On one of my visits to Hisar, this Rohtak family met me and approved me as their potential son-in-law. A few days later, when I was visiting Rohtak to attend a wedding, I met the girl and her family. I was satisfied with the meeting and so were others in family, including my parents. But my father did not want to formalize the relationship without the approval of Dadaji, who happened to be on a pilgrimage to Hardwar. My father told the middleman and the girl's family that the final approval will be given after Dadaji's return.

Meanwhile, someone told the girl's family that we are financially weak. There was a buzz created that our business was not doing well. This family sought fresh assurance from other acquaintances to reaffirm the state of our finances. They were told that our family business had revived. But this assurance did not seem enough for them. We learnt later how deeply sceptical this family was about our business success.

While we were waiting for Dadaji to return, I was busy with my work in Delhi. One day an old family retainer, Dulla, came to meet me from Hisar. He congratulated me on getting engaged. I was a bit surprised. I asked him why he was congratulating me since we were still to formalize the decision.

The mystery was resolved when Dulla told me about the visit of the girl's family to Hisar. It transpired that my prospective in-laws came to Hisar with Har Gopalji to inspect our family property and income. Dulla was asked to show them around our land and mills. Now I realized what had happened. The family had not believed us and had doubted our financial status. Despite our assurances, they had felt the need to physically verify our assets.

This made me very angry. I did not want to have ties with a family that did not trust us. I went to Hisar and cancelled all further talk of marriage with this family. I was especially upset that this was happening though our family had not made any financial demands of the girl's family. My father agreed with my view. We did not want to marry into a family that was more keen on our assets and less on the person their daughter would live with.

Soon after this episode, we received another marriage proposal from a family in Amritsar. This family knew my father through some common friends and distant relatives. They came to meet me at my office in Delhi where I worked and lived. They sat with me in my office for 2-3 hours and watched me work. Soon they were convinced that I was capable of taking care of their daughter. After a few days I met my future wife in the lawns of Birla Mandir in New Delhi. Sushila had been called from Amritsar for our meeting.

This was the first time I was meeting Sushila. And in typical fashion, our families left both of us alone for a few minutes to chat with each other. She hardly spoke despite my prodding; I was a bit flustered, as is natural. But I managed to ask her whether she cooked and what she had studied. She had studied only till class 9. Her father had passed away but her eldest brother was the patriarch of the family. The brothers ran a textile trading business.

Dadaji was a bit sceptical about the proposal. He was not sure whether a cloth trader's family would be suitable for our family. He was not in favour of it and wanted to know how much would they spend on the wedding. I think he wanted a more affluent family, but he was ready to consider them since I had consented after meeting Sushila. To be doubly sure about the family and the prospective bride, in particular, he went to visit them in Amritsar and stayed there for two days. He asked Sushila to cook for him, tested her knowledge of Ramayana and Mahabharata, saw her hands and feet for lucky features.

He returned satisfied. He was more impressed with Sushila than with her family. Dadaji liked her temperament, demeanour and knowledge. For him the person was more important than the family. With his approval our wedding was finalized.

Initially, Sushila lived in Hisar while I was in Delhi. I would only visit on weekends. After two years, when my work had stabilized, I asked her to shift to Delhi with me. I would divide my time between Hisar and Delhi, but had begun focusing more on the business in Delhi. Laxmi was handling the work in Hisar while Jawahar moved to Delhi to support our business.

THE NEXT TWO years were spent in meeting other family commitments. I helped my father in finalizing the marriage of my two brothers and one sister. Since business in Delhi was doing well, I brought Dadaji also to Delhi. I had rented a house in Punjabi Bagh for Rs 550 per month. The mill was in the Lawrence Road food processing zone. By now the family had recognized my position as the emerging family patriarch. I was twenty-four and the eldest among the brothers. Also, my business was stable and I had proved myself by moving out of Hisar and creating a strong business. Now Dadaji and my father were keen that I lead the business growth for the entire family with the support of my brothers.

Jawahar and Laxmi joined me in Delhi. Our youngest brother, Ashok, was in school. With my brothers' help, I could scale up the business and soon our earnings were close to Rs 10 lakhs a year. My father was busy in philanthropic activities. The other grandfathers remained in Hisar and had independent businesses.

5

BETWEEN THE SHEETS

Or how my hair turned grey overnight

A NEW BUSINESS opportunity had emerged for us in 1975. This was also from FCI but did not involve grain processing. That year India saw bumper agriculture production. The Government of India realized that there would be an acute shortage of storage space in the country, and hence directed all government agencies like Central Warehousing Corporation, state marketing federations, civil supplies departments as well as FCI to create additional storage space.

To enable private sector participation, a soft loan scheme was launched. If anyone built a warehouse and rented it to a government agency, the banks would extend loan facility at a cheaper interest rate. To meet the growing need, FCI decided to build additional temporary warehouses using low density polyethylene (LDPE) films and sheets.

The polythene films were fused in a way that they formed a cap or cover of 31 feet by 20 feet with a height of 18 feet. About 1200 bags of grain, weighing 120 tons in all, could be stacked in open space and then be covered with this cap. The grain would remain safe from all weather conditions, including rains.

So far, these caps had a monopoly supplier—Calcutta Commercial Corporation. But the sudden demand meant more vendors were required. This seemed like a good opportunity to me. My friends in FCI told me that the margins of profit were better in this than in processing grain.

Despite the tough competition I would face from Calcutta Commercial Corporation, I decided to bid for the tender to provide extra covers. This was the first time that I was stepping out of the family business of grain trading and processing. I was twenty-five, confident, but had little idea about the challenges, competition and the betrayals I would face.

There was another monopoly involved in the production of raw material (LDPE film). The sheets with the necessary width were manufactured only by Union Carbide in India. There seemed to be a nexus between an official of Union Carbide and Calcutta Commercial Corporation. As a result of this, Union Carbide refused to sell raw sheets to anyone except Calcutta Commercial Corporation. Though it was a cosy arrangement that kept out competition, the scale of the business was small. Only about 200 covers had to be made per year.

In comparison, the new tender was much bigger, with a demand for 1,000 covers.

The earnest money required for the tender was about Rs 70,000. And to service the order, we would need about Rs 15 lakhs of working capital. The financial conditions were tough since Union Carbide would want to be paid for the sheets in advance and would deliver after ten days. The need for liquidity was high.

Despite the difficulty, I was keen to get a slice of this business. I looked for a partner in the foodgrain trade. Rakesh Gupta was young like me and was struggling to get into FCI's grains processing business. He readily agreed to pitch in. He said his family would arrange for the working capital but not the earnest money. In return, he wanted to share the profit and loss equally. I agreed and managed the Rs 70,000 bid money from my existing business. We decided to bid for a thousand covers.

To my surprise, just a day before the bidding process, Rakesh suggested that we split the tender into two. We would each bid for 500 covers each. I could not understand the reason, 'It is because later there will be issues of income tax and other accounting complications,' he said. I accepted, though I was not convinced of the reasoning. I created a company called Delhi Commercial Corporation. Rakesh prepared his bid for 500 covers in his firm's name. Calcutta Commercial Corporation was bidding anyway. FCI accepted all the bids and the allocation went according to plan. Calcutta Commercial Corporation won 1,000 covers. Rakesh and I won 500 covers each.

Now the next step was to source the raw sheets from Union Carbide. But the officials of the company refused to entertain us. We did not enjoy their confidence and from their perspective, we appeared to have no credentials and were without any past experience in plastic fabrication. I went back to FCI executives and requested them to make a recommendation for us.

I finally managed to get through to one Dr Pingle, who was a senior officer designated as Manager QC at FCI. I told him that we had managed to get the tender from FCI, but that Union Carbide was not providing us with the raw material. Dr Pingle was a fair man and after a patient hearing ordered Carbide to supply us with the material.

Just when I managed to overcome this obstacle, I ran straight into another crisis. Rakesh now wanted a larger share of the partnership. Even though we had bid separately, I needed his help for the working capital as he had promised earlier. He threatened to withdraw his support if he did not get a larger share of the covers.

THIS WAS A shock to me. My plans to go beyond grain processing were about to collapse. I was in a crisis and the tension turned my hair grey. I developed my distinctive lock of grey hair on my forehead during this crisis. In about fifteen days, at age twenty-five, I had one-and-a-half inches of grey locks just above my forehead.

Somehow I held myself together. I did not want to give in to Rakesh's demands. But getting the working capital was essential. I had no choice but to seek loan from a bank.

I approached a field officer, Suresh Rastogi, in Bharat Overseas Bank. I had met him on a few occasions earlier. He said I could get a loan of about Rs 5-10 lakhs if my paperwork was complete. But I did not have any collateral to offer.

I had to seek the help of the manager of the branch where Rastogi worked, one Mr Dass. When I met him I discovered his interest in astrology. I got chatting with him about it and praised his knowledge. We could make a personal connection over this informal discussion. Then I asked him for advice and help. He finally agreed to help me since I had won the tender and had assured revenues.

The bank manager had taken a risk on me by accepting my personal guarantee. To support it, the tender document was kept as security against the loan.

After getting money from the bank, I went to my friends and managed to raise the rest of money. The bank loan approval helped me convince others about my business plan. My greying process stopped once I had managed to raise all the money.

Now the challenge was of manufacturing. While I had the money, I did not have any facility or knowledge of fabrication. To add to my troubles, Rakesh wooed away the technical team I had begun to assemble. Rakesh was angry since I had not agreed to his deal. He tried to sabotage my work.

I spoke to another business person to help me since Rakesh had walked out. I had been contacted by a contractor, R.C. Kapila, who was in a similar business. Kapila made plastic bags for packaging and also the covers that FCI was looking for. But his production was limited to ten to twenty per year for some state warehousing corporations. Kapila was keen to take part in the new order. But he did not have the financial or organizational strength to handle large orders. He could offer very little technical support.

During this struggle I had begun to understand the process of making plastic grain covers. I studied how to get raw sheets and how to cut them to the right size and shapes. After cutting, they had to be placed on a table and pressed with hot iron. I had no choice but to go it alone. I hired migrant labour from Rajasthan and started making the covers to specifications.

My next challenge was to ensure that FCI paid me in time. Without the payments, I would not be able to buy fresh raw material, process the sheets and pay the labourers. The loan was not enough to take care of my working capital needs. I had to convince the FCI finance department to clear my payments on a fast track. A friend from my school in Hisar, Kishan Gupta, who was also working in FCI, came to my rescue. He introduced me to officers in the finance department. I had met Kishan by chance on one of my visits to FCI.

We were working round the clock to fabricate the covers from sheets purchased/delivered to us on a given day. I would deliver the covers after a quality check around noon and have them inspected and accepted. Then I would take the bill to the FCI office, and get it processed by the evening. The next day, the finance department would issue me the cheque. It was not easy for me to get this done. To get payment within a day was a tough task—FCI normally took about ten to fifteen days to process a cheque—and I had to offer my pound of flesh. The finance department would ask for small favours and I would oblige.

I managed to maintain a constant flow of money, which

allowed me to fulfil my contractual obligation of supplies in time to avoid penalties for late supplies. We could maintain a cycle of three days from the purchase of raw material to converting it into cash for the next purchase. Here, both my brothers, especially Laxmi, played an important role. Laxmi would stay at the place of fabrication for 24 hours. He would not sleep for more than 2-3 hours.

Meanwhile, a senior official at Union Carbide took a liking to me as he thought I was very industrious. He issued orders that I be supplied sheets based on the payment I was making. Normally, the company would sell 10-12 tons of sheets in one instalment. But I would never have enough money to pay for the entire lot. So I was able to buy 2 tons at a time. Buying small lots of sheets helped me meet the manufacturing and supply deadlines. I earned a profit of Rs 3-4 lakhs from making those plastic covers for FCI.

Somehow I kept making friends along the way. Some because they were taking favours from me, others because they respected my passion.

After this success, I bid for more covers. I went on to make about 25,000 grain covers for FCI between 1975 and 1978. Such was the need that even Union Carbide could not supply enough sheets. So FCI had to import the raw sheets from Italy. The harvest was so high in those years that the government was ready to help anyone who could create storage capacity.

FCI imported raw sheets in Delhi, Mumbai and Kolkata separately to meet regional storage needs. But they also had to issue fresh tenders for their fabrication and conversion into covers. I won the bid in Delhi for the north region.

This work did not require any machinery as such. It required about 300 labourers and a large space for heat-pressing/welding the sheets. When the volume grew, I had work going on round the clock at the site with the labour working in shifts. The raw material would arrive during the day. We would work all night to convert it into covers. We would get it packed and delivered to FCI by morning. My brothers Laxmi and Jawahar worked almost nonstop to maintain this punishing pace.

Our efficiency allowed us to process high volumes. We bid for the tenders in Kolkata as well. I also travelled to Mumbai and bid for the tender there. I rented a large commercial space/ godown in Kolkata. In Mumbai, it was tough to find commercial space without *pagdi* (an advance equal to a year's rent). My brother-in-law, Basant Badgamia, helped me find space in Bhandup at low cost. But the owner demanded Rs 5 lakhs as *pagdi*. I sat with him in the evening for a drink. I chatted with him in Punjabi for many hours. We ended up chatting like old buddies. He started trusting me and waived the deposit of *pagdi* for me. Developing a strong connect with people has always helped me.

Rakesh did not do well in the next set of tenders. I believe that someone who cheats or does not keep his commitment does not do well in life, especially in the long run. He came to me later and apologized. He wanted to make up and be partners again. 'It is over, Rakesh sahib,' I told him. 'You almost bankrupted me. I can't work with you again.'

Meanwhile, I got a call from Dr Pingle of FCI. He asked us to manufacture covers on an experimental basis with a different and new combination of plastic and fabrics made from HDPE (high-density polyethylene). He said we should try to combine the two by laminating high-density polyethylene fabric with low-density polyethylene plastic. Also, earlier, the fumigation cover was made with more expensive rubberized cloth. Dr Pingle wanted to experiment with cheaper and more efficient material. These covers would have to be airtight so that the fumigants (chemicals) put inside them would be effective and kill the insects in the grain storage.

I started looking for a lamination factory. I found one but the owners, Purushottam Gupta and Balkishan Goel, would not rent it out for less than a two-year period. I wanted it only for three months to experiment with the new combination. I had no choice but to commit for two years and hoped that the experiment would succeed and that we would get repeat orders. Then I looked for manufacturers who made the HDPE-woven fabric. We found some units and managed to make about 2,000 fumigation covers for FCI.

But this project was not successful. The fumigants were chemicals that turned into gas once they came in contact with air. They had to be kept in airtight conditions within the covers for effective fumigation. If the gas leaked, the fumigation would be wasted. We were stitching together the fumigation covers and plugging the stitches with a liquid rubber material but it still did not make them airtight.

Here I must narrate an anecdote where I had to play a little trick to get past a difficulty. For the experimental covers there was a total period of delivery and even a day's delay beyond the stipulated period would have done two things:

• FCI depots would not accept delivery after the deadline expired.
• The security deposit of Rs 70,000 would have been forfeited, making it a double whammy for me.

Two days before the deadline was to expire, I discovered that the suppliers of HDPE fabric had let us down and not delivered the desired quantity.

I could not think of a solution to the problem. I had yet to deliver one more truckload of a hundred covers. I had gone to the FCI warehouse for delivery with Chhagan Lal (our accountant-cum-delivery man), when suddenly I thought of a solution. I had to bend the rules and take a big risk. If this were to come to the knowledge of FCI officers, I would have lost all my credibility.

I found out where these hundred covers were to be sent. Then I got a cousin of mine, who was a transporter, to bid for the transportation of these covers. FCI would put out daily tenders for transportation. My cousin quoted cheap freight charges and got the bid for the delivery of these covers at an FCI depot in Rajasthan. Once he got the covers on his truck, I brought them back to our factory. I changed the serial numbers, and repacked them and submitted them again. This round-tripping bought me some time. In a few days, I got the extra raw material that I was waiting for. After this all the covers were delivered to the destination in Rajasthan and no one noticed a few days' delay in transportation between Delhi and Rajasthan.

I FACED A near-disaster in a deal again because of trusting someone. Our accountant, Chaggan Lal, brought in a young person to assist him. Our work had increased and Chaggan couldn't cope with it.

This young new accountant was one Om Prakash who started well. Our entire staff was three persons—two accountant-cum-office assistants and one peon—in addition to us three brothers—Laxmi, Jawahar and me. The three of us handled different functions. Sometime in 1976, when I was preparing an FCI tender for plastic covers, I noticed something fishy. Om Prakash started getting curious about the tender details a day before the deadline. His job was to prepare the final bid papers. He began asking about the final rate to be typed in the bid document. I ticked him off, saying that there was more than a day to go for the deadline and it was too early to finalize. But he kept pestering me. I felt uneasy.

Still, I gave him the rates and asked him to finalize the bid and seal it. I gave the envelope to Chaggan and asked him to put it in the bid box in FCI by 11 a.m. the next day. The deadline was 12 noon. Both looked delighted.

Dadaji had taught me one important skill. He had told me to look into the eyes of people to assess what they were thinking. Dadaji would initiate a conversation by talking about everything under the sun, except the business at hand. His philosophy was simple. The more you chatted with someone, the better you could know him.

I had applied this to Om Prakash and Chaggan Lal. Something about their sense of excitement made me suspicious. I told them that the tender bids would be opened at 3 p.m. the next day. And asked them to attend the bid opening as I had some other important tasks to handle.

They promised to take care of everything. I said that the bid rate was enough to take care of our profit, and the competition was only with my old rival, Calcutta Commercial Corporation.

They left for the day and so did I. But later that evening I returned to the office and prepared a new letter. I mentioned my earlier bid document and wrote that we had reconsidered our bid rate. I changed the figure in the bid to a lower rate and requested that the new figure be considered as final. Then I went to FCI directly and called a clerk who knew me. I requested him to drop this in the bid box after 11 a.m. but before 12 noon.

The next day, when the bids were opened, there was a surprise. Apart from Calcutta Commercial Corporation and me, a new company had made a bid. Their bid was lower than our first bid. Then the penny dropped. Our accountants (Chaggan Lal and Om Prakash) were working with this new group without my knowledge. They had used my figures and quoted below it. But they did not know that I had put in a revised bid, which was even lower than the new company's offer. The new company was surprised at the turn of events and my accountants were in shock. I won the bid and later confronted the accountants.

My brothers and I were so enraged that we physically thrashed the accountants. Chaggan Lal had worked with us for twelve years. He admitted to helping the new company. He said that Om Prakash had convinced him to give all the details to the new company. They explained the entire bid process to the new company in return for a big fee plus a share of the profits.

I sacked Om Prakash but kept Chaggan Lal, for this was his first mistake in twelve years. Om Prakash had corrupted him. Chaggan Lal would stay with us till he retired, and even worked for the launch of Essel World in 1990.

6

DIVERSIFYING INTO PACKAGING

Acchey din arrive as I spread my wings

I WAS SETTLING in nicely into life in the big city. I bought a second-hand Fiat car around 1974. This was an old model, where the front and rear doors had hinges on the central pillar. So the front door opened facing the engine. It was a cute and unique model. We lived in Punjabi Bagh and enjoyed driving around the city. We would go eating out to places like Moti Mahal in Darya Ganj and to Pandara Road restaurants. Those days some of the restaurants had ghazal performances in the evenings.

This was a fairly stable and enriching period in my life. Our family had tasted success after a long while. I was twenty-four, married and could enjoy material comforts. The days of sleeping in a mill were over. I felt that we could take on the risk of a new business and be successful.

My horizons had widened. From being a small-town trader, I was flying off to cities like Mumbai and Kolkata. I was the first person in the family to take a flight at age twenty-one. It was a day trip. I had to rush for a tender so I took an Indian Airlines flight. A one-way ticket was about Rs 240. A ticket in an air-conditioned railway coach would have cost about Rs 80.

Now that the FCI business was stable, I began to think of something new. The FCI work was comfortable but unpredictable. We lived from bid to bid. I was keen to start a business with more certainty. Moreover, I wanted to diversify so that we were not dependent on one business. My grandfather and the family were totally dependent on the grain and agriculture trade. I wanted to ensure that we had enough options so that we would not collapse. I was keen to diversify and de-risk our earnings with a few small steps.

Meanwhile, the factory I had rented for the fumigation covers was lying idle. The experiment had not succeeded but I was locked into a contract of renting the factory for two years. We wondered what to do with the space. The factory owner, Purushottamji, suggested that we start packaging work for pharmaceutical companies. The key work was to laminate paper with polyethylene, and offer it to these companies. Those days, medicine tablets would be packed in white glazed paper laminated with plastic.

This idea appealed to us. It was a new business but did not look complicated. We launched a company called Lamina Packers. I sourced the paper from the distributor of a paper mill whose office was at Chawri Bazaar in Delhi. I kept a salesman, too, for selling this material. It was perhaps the first salesperson I hired. We started getting small orders from pharma companies in the Delhi region. But it did not work out very well because we were making plain and unprinted laminated paper. Pharma companies wanted packaging paper that had text printed on it. If they bought plain sheets, they needed someone else to print the medicine details. To meet this need, I had to invest in a rotogravure-printing machine. I had to rent another shed where we placed a second-hand printing machine for the work. This would be the beginning of our entry into the packaging business. Of course, I did not know then that one day we would become a global player in this field.

Delhi had very limited scope as most of the big pharmaceutical companies were in and around Mumbai. I decided to open a sales office for Lamina Packers in Mumbai, too. We rented the office in the industrial belt in Mumbai to be locally present. One of my friends, C.K. Jhunjhunwala, offered to give me his office space without charge. But I insisted on paying a token rent of Rs 500. I did not want to take unnecessary obligation. Seeking a favour from a friend is one thing, but exploiting a friendship is another.

I was very excited about opening an office in Mumbai. I actually took almost fifty friends and relatives from Delhi to Mumbai to inaugurate our tiny sales office! Now when I look back, I realize how silly it was.

At another level, though, it was a big achievement for me as it was my first expansion outside the Delhi-Hisar area. This office helped us penetrate into the market for supplying large companies. Our plant in Delhi had enough orders to run on full capacity. I felt blessed. The packaging business began only to utilize an idle factory. But it became a profitable diversification for me.

The factory itself was the result of an experiment that we had done for Dr Pingle. Sadly, I lost touch with him when he shifted out of India. I heard later that Dr Pingle had joined a United Nations organization to work on a larger scale. He is one of the few people who helped me but with whom I could not stay in touch.

We graduated from paper laminations to plastic and related laminations for different packaging materials. The effort opened up new opportunities and shaped our future entry into the big league of packaging.

Along with Lamina Packers, we considered other businesses, too. We started work in the fibreglass moulding business. Fibreglass can be moulded into various products such as bathtubs, chairs, helmets and so on. At that time there was only one company, Fibreglass Pilkington, that made the raw material. Another company that started the manufacture was looking for partners who could mould the products. The new company encouraged us by offering to pay 10 per cent of the capital expenditure from their development funds if we made new products.

This seemed like a good arrangement. We zeroed in on ceiling fans as a product. We set up high-pressure moulding machines to make top and bottom covers of ceiling fans. We worked to create this new category but it turned out to be a bad idea. Fibreglass is a heat absorbing material and was not suited for ceiling fans. The copper coils inside the fan generated heat, which was absorbed and not transmitted outside by the fibreglass. As a result, the moulded cover of the fan lost its shape and affected the functioning of the fan.

By 1977-78, the grain processing and other business with FCI was declining. It had become too competitive. I wanted to stay in a business where I was number one or a strong number two. It was becoming tougher for me to maintain my position in FCI. From just three to four, the number of bidders for FCI work kept increasing. Undercutting beyond a point was difficult as it would involve compromising on the quality. In all the work that I had done, I had not compromised on quality. By 1978, the FCI work was over.

7

POLE-VAULTING TO A
HIGHER PLANE

I take a risk but get cheated...again!

MY LIFE WAS about to take a new turn. While we were winding down our FCI work, I came back in touch with my old business associate C. Lal, whose accountant had cheated me during my first stint in Delhi. Lal owned a regional private airline called Jamair in eastern India. Lal also owned some flour mills. His office was in Akash Deep building on Barakhamba Road in New Delhi.

Some time in 1977, Lal phoned me. This was after a gap of four to five years. I went to meet him in his office. He was a Marwari businessman with roots in Bengal and Assam. But he spoke in the traditional way. *'Kis tarah chaal rayho hai, dhando achho hai main sunyo tere paas?'* He asked me about my business. I replied that I was doing very well. *'Paise peese ki sahuliyat hai kei?'* He asked if I had some spare money to lend him.

I replied in the affirmative but asked him how much he needed. He then asked directly. 'I need about Rs 10 lakhs for two months. Can you give it today?'

I was shocked by this proposition. And confused, too. At one level, I was angry since this person had been a party to cheating me and causing me much humiliation before my family. At another level, I felt powerful. I felt honoured that a big industry leader like Lal was seeking financial help from a small-time trader and manufacturer like me. I felt that I was being seen as an established businessman.

I told him that I would arrange the money in a couple of days.

I returned home and told Jawahar that Lal wanted money. Jawahar warned me not to trust him. He reminded me that Lal belonged to the Maheshwari community of traders, while we were Agarwals. In Hindi, the word Maheshwari is spelt with characters that have a loop in every letter. The lore was that those who traded with them, were certain to be entangled in one or the other of their many loops.

But I was not convinced. I decided to help Lal. I told Jawahar that we had about Rs 50 lakhs worth of total cash and assets. We could spare about Rs 10 lakhs. So despite the objections I went over and gave him a cheque of Rs 10 lakhs. I got busy with my work and about three months passed. I started asking him to return the money. Initially, he said he needed some more time. But after a few weeks, he stopped taking my calls. Over six months passed and there was no sign of his returning the money. I did not have any agreement on the loan apart from the cheque payment. One day I went to his office first thing in the morning; he had not yet arrived. When he heard that I was waiting for him, he did not even turn up at work.

I turned to his manager, an elderly person called Bajaj, and asked him what the problem was. He indicated that times were tough for Lal and that he was in a deep financial mess. I asked Bajaj to inform Lal that I would keep coming to the office until my money was returned. And if Lal did not come to office, I would go to his home in Greater Kailash.

I also met B.R. Chopra, who had worked with Lal and me

earlier. Chopra facilitated meetings in government departments and had helped Lal and me get some tenders. Chopra had stopped working with Lal but confirmed what Bajaj had said. It appeared now that I had got myself into trouble.

Chopra himself had an interesting story. At one point of time (as I was told), he was the chauffeur of Congress leader Jagjivan Ram. He developed political connections by being close to him. He gave up his job as a driver and used his connections to get things done in Delhi. He had also partnered with Lal for some project. Chopra was reportedly close to O.P. Arora, private secretary to Jagjivan Ram, and R.K. Dhawan, personal secretary to Indira Gandhi.

Lal finally met me when he realized that that he could not get rid of me so easily. However, instead of returning my money, he made a fresh business offer. He said that his company had bagged a big order to make telecom and telegraph poles for the Post and Telegraph department. Lal already had a factory to make poles in Faridabad. The value of the order was large and if we succeeded in making the supplies we could get more orders worth tens of crores.

I asked him what the margin of profit was. He said it was 15-20 per cent. This worked out to about Rs 3-4 crores. The work required a minimum investment of Rs 1 crore.

Logically, there was no reason for me to accept the offer. Lal was not to be trusted. He had cheated me once and was trying to succeed a second time. Moreover, my brothers were against our doing business with him. Thirdly, we did not have assets worth Rs 1 crore.

Though this was a risky proposition, I was attracted to the offer. I was bored of my current work and was looking for a challenge. My confidence level had increased as I had been able to do well in Delhi. The profits promised in the venture were several times higher than what I had achieved so far. From earning in lakhs, I would be earning in crores. And the prospect of a totally new business was exciting.

Also, I felt that we had the credibility to raise the Rs 1 to 2 crore required for investment.

I thought about it and finally agreed to accept Lal's offer. But this time I put in three conditions to protect myself. One, I would own 50 per cent of the business. Two, I would manage and control the pole-manufacturing unit. And three, I would control the finances for this company and the venture. My signature would be a must for all transactions. I told him that his past behaviour did not allow me to trust him with money. I said that apart from him, I would be the co-signatory for issuing cheques. Lal was clearly desperate for help. He agreed to all my conditions. And we prepared a draft agreement on the spot.

When I returned and told my brothers, they were furious with me, as was to be expected. They were shocked that not only was I not getting our Rs 10 lakhs back, but that I had promised an even larger investment to him! 'Our assets are only about Rs 40 lakhs. Where will we get an additional Rs 50-100 lakhs for the venture?' they asked. Moreover, our assets were invested and not liquid. My brothers registered their objection but I explained the advantages. I managed to convince them that this was the only way to get our money back. Importantly, we were also getting a chance to enter a new business that could put us into a higher profit orbit. The final decision was mine and I felt convinced.

Raising money was going to be tough for us but manageable. We were in the middle of several tenders and much of our capital was blocked in these. Each venture required some money to be blocked as bank guarantee. This sum is released only after the contract is completely delivered. There were also working capital needs for each of our mills.

I was confident about the project as I was going to directly handle the manufacturing of the poles. We then set about raising money. I borrowed some money and took some out of our existing projects.

We now needed someone to manage this project. I reached out to a friend, Subhash Grover, and requested him to leave his job to take charge of the project. The company handling the pole project was called Suraj Lamp, and belonged to the Lals. They were losing a lot of money in this venture due to inefficiency and a lack of working capital.

I stumbled upon the real crisis in Lal's company when I started dealing with his banker. The payment from the telecom department was made to Lal's account in State Bank of Saurashtra. But the bank would deduct 10-15 per cent before passing on the rest to us. I met the manager to find out why. He explained that Lal owed the bank Rs 7 crores. And the bank was deducting the sum to recoup its money. This was a shock to me. This explained Lal's desperation to get me in as a partner.

I was upset and told the bank manager that the agreement I had with Lal did not allow anyone to withdraw money without my permission.

The bank did not recognize my agreement with Lal. The bank had the first lien on all income of Lal's company. So I did a deal with the bank. I would personally inform the bank about the profit of the company. And the bank would keep a percentage. I was their only hope for the recovery of the dues. I explained to the bank manager that if my partnership failed, then they would never be able to recover the money from Lal. But if the venture succeeded, Lal would have the money to repay the bank. After this discussion, the bank supported me.

We took charge of the project. We saw that the company was losing money because of inefficient management. The company was buying raw materials from traders at a high price. We decided to cut out the traders and buy from the manufacturers directly. It saved some costs. We also stopped leakage of material from the factory. By taking more such steps, we could reduce the costs and improve the manufacturing process. The company started running smoothly and with higher profit than what Lal had promised me. We not only fulfilled that contract in time but also got more orders. The P&T department found us a more reliable and efficient supplier than Lal.

The bankers were happy as they recovered close to Rs 5 crores of debt. And Lal and his family were happy that we had managed to turn around the company. He and his sons could reduce their debt and earn without too much of effort. Sadly, they did not appreciate the profits they were earning without working for it. Their natural instinct was to try and keep the profit with themselves, though it was generated by my efforts. Lal had two younger brothers who were part of the business. He called me to his house one day and began yet another drama. He said his brothers were upset that Lal was not the sole signatory in the company for banking and for making any other commitments.

Lal quoted his brothers saying, *'Bhaiji, thare dastkhat ke uppar Subhash ka dastkhat howe hai, ye to aapni beizati hai.'* They felt insulted, as my signature was more important than Lal's.

I laughed and told him that this issue was being raised as the business was not only making profits but was also bigger than anyone anticipated. I reminded him that I had revived the business with my brothers through sheer hard work and application of mind. His brothers had no role in its success and therefore had no business commenting on it. Lal agreed but did not say anything.

I knew what was going on in his mind. 'Do you want to break this partnership?' I asked. He felt embarrassed but meekly said 'yes'. Lal was under pressure from his brothers but his wife thought differently. She said, *'Maine to inko kaha hai ke Subhash ka bhag kaam kar raha hai, humein use alag nahin karna chahiye.'* She said that my good luck and fortune had brought them profits and that the partnership should not be broken.

I had to take a quick decision. Lal and his brothers were not to be trusted. I was ahead of the game here and had made decent profits. Staying on with them would only hurt me in some way. I took the decision on the spot to end the partnership. I said it was time to part ways. I told him that I would now leave the business and that he should buy me out. I left his house after saying this.

We ended the partnership but Lal did not settle our accounts. The original investment of Rs 1 crore in the business was already received by me, apart from a couple of crores of profits. But in total we did not get Rs 5-6 crores from Lal. Despite that, we earned more than Rs 5 crores from the poles business.

The decision to take charge of this business was a great learning experience for us. We ran it from 1978 to 1980. Morever, I had achieved what I had anticipated. From earning in lakhs, I had taken our business turnover into crores. Despite my terrible experience with Lal, I had emerged stronger and had a better balance sheet.

After I left the patnership, Lal and his brothers could not run the company as efficiently as us. They started making losses again.

As I look back, I find that people who cheated me did not do well in life. My uncle Ghisa Ram had taken away our family assets. He suffered later because of the accident. Rakesh Gupta had let us down in the polythene covers business. Later, he developed serious differences with his brothers and the family split. They lost so much money that they were bankrupt.

I was told by many people that I brought my luck to my

partners. My fortune supported them as long as I was with them. Once I parted ways, their luck did not hold out. This could be sheer coincidence, of course. But finally, it happened to the Lals, too. Recently, his youngest son met me and said, *'Bhaiji, thaare saga mahri takdir bhi chali gai.'* They lost their fortunes and luck when I left the business.

Throughout this period Dadaji kept a watch on my business. He was over sixty and leading a retired life. He stayed with us at our house. Sometimes, he would drop in at our office for a chat with my brother and other people in our company. He would always tell me that I was *bhola*, too innocent and trusting, and that I would get cheated by many people.

I did not contest that, and agreed that I had been made a fool of several times. But I did remind him about the distance I had covered. From 1971 to 1980, our balance sheet had improved every year. Our turnover was close to Rs 15 crores and our profits were more than Rs 50 lakhs per year. Our profit had grown every year, and the business was still growing steadily.

ASSESSING A PERSON correctly has remained an issue for me. Most of the time I am correct in my assessment but I do occasionally get it wrong. And when I do, I get it totally wrong. To some extent, Dadaji was right about me. My first instinct towards someone is always of faith and trust. If I am proven wrong, then so be it.

Ironically, incidents with people like Rakesh Gupta and Lal gave me confidence. I survived such shocks and betrayals. Also, I became smarter and learnt to cut a better deal for myself. By setting the three conditions in my agreement with Lal, I did not suffer losses. I might not have received all my dues, but I was still in profit. Most importantly, I learnt that I could be successful in a new business even if I did not have any experience in it.

Of course, not everything I tried was successful. After the poles business, I invested about Rs 25 lakhs in a hand-tool factory. By then my youngest brother Ashok had grown up. I asked him to start handling this work. He worked hard in the factory for a while but it did not work for us. The sector was too competitive and tough to survive in. The hand-tool manufacturing was reserved for the small-scale sector and depended a lot on government subsidies and tax exemptions. Most players in the sector were making money from exemptions but not doing any real business. Finally, we sold the factory and the business.

During this period the Central Bureau of Investigation (CBI) also raided us. Without any warning the CBI landed up at our office and home. I rushed to ask them what the charge was. I was told that a manager of State Bank of India had been arrested for holding disproportionate assets. And our name had figured in his list of clients.

The CBI suspected that I had worked with that manager to defraud the bank. We were confused since we had nothing to do with the fraud. We tried to get more information about this case. We discovered that this manager had issued a bank guarantee to us for an old FCI tender for fabrication of LDPE covers for storage of grains.

The CBI told me this manager had issued the guarantee without taking clearance from his superiors, and that the entire transaction was not recorded in the bank's books.

The CBI thought that we had paid a bribe to the manager for a fake bank guarantee. Hence this raid on us. They searched everywhere to find evidence, but they did not find any. However, after a few days, I received a summons to personally appear at the economic offences wing of CBI's Delhi branch.

I suffered for four days and four nights. The officers of the economic offences wing grilled me for many hours every day.

They would begin in the morning and question me till midnight. The same question would be asked three or four times in different ways. Different officers would come and question me time and again.

When they did not find any evidence they resorted to underhand tricks. One officer found a few adult movies in the house, which I had bought for my private viewing. They threatened to file a case against me for keeping obscene material. But soon they gave up, as private viewing of such material is not an offence. They were trying to use the movies to scare me into a false confession.

We were fortunate because there was enough proof to show that the guarantee issued to us was genuine. There was confirmation by the same bank manager to show that the guarantee money was returned to the bank directly by FCI.

While the guarantee was genuine, the manager had not taken prior permission from his supervisors. He had asked us for favours in the name of his seniors.

By the time this matter was closed, a few CBI officers became my friends. They were convinced of my innocence. I maintained my relationship with them by meeting them occasionally.

8

THE RUSSIANS ARE COMING

A Swamiji, the Gandhis, and a very big deal

BY NOW I was well established in Delhi. While working on the telecom poles project, I had understood how Central government departments worked. I met many persons who were in the government or were politically connected. Delhi has a culture of encouraging intermediaries. Most of them have worked in government or assisted someone in the government. All of them claim to be well-connected, with a deep understanding of how files moved in various ministries.

As my circle of contacts increased, I met one Mr Tripathi who claimed to have worked with Sanjay Gandhi, son of Indira Gandhi.

How I met Tripathi is another interesting story. I met a government relations intermediary, one Mr Sharma, at the office of Purushottamji, who was the owner of the lamination plant we had taken on lease. Sharma would visit very often. I was told by Purushottamji that he was an officer in Research and Analysis Wing (RAW). Bit of a greenhorn, I did not know then what RAW did. Purushottamji told me it was something like a CID arm of the government.

On one occasion I asked Sharma for help. A friend in FCI,

M.R. Behl, wanted to take voluntary retirement and join the firm of another associate of ours, Anil Chanana, as working partner. The problem was that Behl's boss was not allowing him to leave; he was not accepting his resignation. I shared this problem with Sharma, who turned out to be very effective. I had made this request in the evening and the next day was a holiday. But Behl's boss signed the relieving order despite it being a holiday and sent the document to him the same day.

Sharma shot up in our esteem after this episode. I was happy to befriend him. Sometimes he would borrow money from me. I would lend him my car. It was Sharma who introduced me to Tripathi. Since I was fairly successful as a businessman and was seen as a resourceful person, Sharma decided to earn some brownie points with Tripathi. He also wanted to impress me with his contacts. It worked both ways for him.

Sharma later got caught in a case of fraud and committed suicide. But my association with him led me to Tripathi, which created connections that would propel me towards people who ruled the country.

Those days Indira Gandhi was not in power. She had lost the elections in 1977 and the Janata Party was in power. She lived in a private house in Nizammudin in New Delhi with her family. The Shah Commission hearings were going on those days and the Gandhis were busy deposing before it.

Tripathi started calling me for help in small matters. Sometimes he requested me to arrange for a ticket for the Gandhis. Sometimes he would borrow a few thousands rupees from me. I would also at times visit him at the Gandhi residence. During some of these visits I met Sanjay Gandhi. I must have met Sanjay six to eight times in all. I would always ask him if there was anything I could do for him. He would acknowledge this without saying much, except that I was helping already. If anything was required, he said, Tripathi would contact me. I never asked for any favour in return. So perhaps they liked that about me.

Those days it was easy to meet the Gandhi family since they were not in power. I stayed in touch with the family and with Tripathi for a couple of years, while they were in the Opposition. One day my associate Anil Chanana came to me for help. He had some problem with the CBI, and asked me to intervene. I was surprised and asked him why he thought I could help. Anil said that some CBI officials were talking about me when he was at their office. These were my friends from the days when they were investigating the bank manager.

I took him to my friends so that the issue could be resolved. On the way back from this meeting, Anil asked me if I had other such influential friends. I said that I had some friends who had political connections. I did not mention that I knew Sanjay Gandhi, too.

About four to six months after this episode, elections were held and Indira Gandhi returned to power in 1980. But there was bad news soon. Sanjay Gandhi died in an air crash the same year. When the news came, I cried. I was upset, as were millions of Indians who were fond of the Gandhi family.

After that I stopped visiting the family as it was now difficult to meet them. My connection with the family almost ended with this accident as Tripathi also stopped calling me; I never saw or met him despite my efforts to contact him. I heard later that he had passed away.

ANIL CHANANA ASKED me again about my contacts after a few months. By then I was also looking for new work since since I had ended my telecom pole partnership with C. Lal.

The Lamina Packers work was still on. Our work in FCI for covers had stopped but a little bit of the grain processing business remained. After earning substantially in the telecom pole deal, I did not find the grain business exciting or lucrative enough. I was keen to start another ambitious project.

Chanana had some new business ideas. He told me that trade with the USSR was picking up and that it was profitable to do business with that country. When I probed him for more, he was not forthcoming. I told him that I would not help him unless I had the details. I was not willing to use my connections unless I knew what the business was.

Then he opened up and talked about exports of basmati rice to USSR. India had a pact for rupee trade with USSR and rice was in great demand. The government was organizing the exports through its canalizing agencies and private suppliers. 'Subhash, you will not believe the kind of profit there is in rice export to the USSR,' he said.

But Chanana also said that a rice export order from the Soviets will require high-level contacts. He said such contracts would not be possible with access only to the commerce or foreign minister. I asked him how much profit there was in this business. 'You forget about figures and numbers. There is so much profit that you will not be able to count it,' Chanana said.

I asked him how much investment was required. About Rs 20-30 crores of working capital, he answered. I told him I did not have so much to invest. Chanana had thought this through. He knew what he had and what he wanted from me. 'Arranging investment is my job and getting the export order is your job,' he said. In return, he offered to make me an equal partner with half the share in profits. 'You get the contract, I will organize the money,' he said.

This was a semi-serious conversation with Anil. He didn't really believe I could get the job done. He was just tossing around this idea with me and some of his other friends. But it stuck with me and I started to work on it. I realized that if this business required a working capital of Rs 30 crores, the returns would be in multiple figures. This looked like a good opportunity for me. It fitted into my plans since my telecom pole business was over.

I began conversations with my friends and contacts who had political connections. There were some MPs from Haryana who I had met over the years. Some I had met during my visits to the Gandhi house. Bhajan Lal, who would go on to become the chief minister of Haryana, was a family friend. His family also hailed from my village, Mandi Adampur.

Some of these friends tried to help me but none could succeed. I remained persistent and invested my time, energy and money on this until one day I happened to meet Rajinder Mittal, a friend and distant relative from Hisar, at a wedding. He said, 'If you have any major work to be done at any level in the government, I can help.' The Mittal family had low credibility in our society circles, but since I had not succeeded I thought this could be the break I was waiting for.

He spoke with such confidence that I was prompted to ask him, 'Who is your connection?' With some prodding he told me that he was close to Swami Dhirendra Brahmachari. Though I had also heard about the swami, I was not sure if Mittal was boasting or whether he really knew him. I said that if he could help me meet Swamiji, I would reveal my business idea in his presence. Mittal agreed and took me to meet Dhirendra Brahmachari. I told Swamiji I wanted to export rice to the USSR. He asked me who would be able to get the contract. I told him that Rajiv Gandhi could get this done, apart from the prime minister herself.

After three to four meetings and lengthy discussions, Swamiji agreed to help me and said he would arrange a meeting with Rajiv Gandhi. I got a call and was given an appointment with Rajivji at 2 A, Motilal Nehru Marg. This was either towards the end of 1980 or the beginning of 1981. This happened to be the office of Rajiv Gandhi. I was excited yet anxious to find out what was in store for me there. I was asked to meet Vijay Dhar, who had been appointed to help Rajiv in managing the political establishment and to help him understand the art of realpolitik. Vijay Dhar was Rajiv's Man Friday of sorts and known to be his friend as well. His father was D.P. Dhar, who had been India's ambassador in the Soviet Union. He had brought the Soviets closer to Indira Gandhi.

Vijay Dhar wanted to know all the details of the business and how I had met Swamiji. He took my number and gave me his and explained the mode and process of communication.

'We will certainly help you. I have to discuss this matter. You should be in touch with Swamiji,' he said. That was the end of my meeting. The entire meeting had lasted ten minutes.

About ten days passed. No word.

Then I got a call from Swamiji, asking me to see him. This time he was keen to know the details of the business. I gave him details of prices, total annual purchase by the Soviets, who they were sourcing from and the profit margin in it. I had come well prepared since Anil Chanana had briefed me about this. Finally, Swamiji asked me what the share of profit would be for him. After detailed and lengthy negotiations, the profit-sharing arrangement was finalized. My friend Mittal was mediating, as he also had to get something for helping in the deal. Swamiji told me I would be required to pay some amount in advance towards profits. We agreed on an amount of Rs 50 lakhs. He also asked me to go and meet Vijay Dhar again after fixing time with him directly.

I went for a meeting at the same place; this time I saw two young personal assistants at the reception. One of them was V. George. There were three rooms in a row. One was for Vijay Dhar, another for Arun Nehru, while the room in the centre was that of Rajiv Gandhi. I waited for more than an hour, watching the activities. Finally, Vijay Dhar appeared from Rajiv's room and Rajiv followed him.

Dhar introduced me as the person sent by Swamiji. Rajiv asked my name and I introduced myself. He was looking at me as if he had seen me earlier. His eyes were trying to place me. I realized this and quickly offered, 'Sir, I used to come to your house to meet Tripathi and Sanjayji when the Shah Commission enquiry was going on.'

'Oh! So you are that Subhash,' he said and smiled. 'We will certainly help you,' he said and walked back to his room.

I followed Dhar to his office. 'Have you talked to Swamiji?' he asked.

'Yes sir,' I said. He said that the bilateral trade agreement between India and the USSR was likely to be signed in the next few weeks. After the agreement was finalized, the USSR's trading company, Exportkhleb, which bought rice and other foodgrains, would send its team to Delhi. The team would meet all prospective exporters to assess their business capabilities.

Dhar said, 'You should be prepared to go through the interview with the Exportkhleb team when they visit India, and if they are satisfied that you can deliver the right quality, in time they will buy from you; otherwise you will be rejected. We can only recommend and not force them.'

He further added, 'I have my serious doubts that you can fulfil the obligation as you have never exported rice or any other commodity.' He remained stone-faced despite my assuring him about our capabilities. He extended his hand to shake mine, indicating the end of the meeting.

Now my priority was to arrange Rs 50 lakhs for Swamiji. I was not too concerned since this was Anil Chanana's responsibility. I fixed up to meet him the next day at his Naya Bazaar office. His partner, Lajpat Rai, was also present at the meeting. They were pleasantly surprised to hear that I had done all the work and that we could be awarded the export order of basmati rice. But I did not share the details of people I had met and spoken to. I felt it was not desirable and I was in any case supposed to keep their names away from all these deals. I had to be discreet, as expected in a situation like this.

I told Chanana that I had made the right contacts and that the order had been assured to me. I asked him to trust me. But he did not seem convinced. More than him, his partner was not ready to believe that I was resourceful enough to get the order when many other businessmen had failed. Rice trading was the monopoly of a few exporters from Bombay who were dealing with the USSR and Exportkhleb.

Chanana and his partner asked me to wait in another room while they discussed the proposal. Finally, they announced that they were not ready to risk Rs 50 lakhs in advance, because if the business did not materialize, their money would be lost.

I agreed that if the deal did not materialize, it would be impossible to get a refund from the swami. It was a business risk, but Chanana and his partners were not willing to take that risk.

'You bring the contract from them and then we will arrange all the funds required,' Chanana said. He wanted to play safe. I tried to convince them the money was safe and that very important people were helping us. But I failed to satisfy them. I shook hands and came out of his office and started walking down the stairs, when I suddenly realized that I had come out without clarifying the status of our partnership. If they were not sharing the risk at this stage then I could not remain committed as a 50 per cent partner.

I climbed back up to his office and said, 'Anil, we are no more partners in this rice export deal. If I succeed we will have a fresh discussion at that time and the deal structure may be different.' They both readily agreed, as they were sure that this was not possible, and internally they were happy to have saved their money.

But I came out a worried man: where would I get Rs 50 lakhs? It was a big amount for me despite my success in business. Most of our funds were either spent or paid to the creditors. Much had been spent on weddings and purchase of some properties. And some cash was invested in the day-to-day business. I was still running small businesses and did not have the liquidity and working capital to fund a Rs 50 lakhs deal.

Once again my family helped me. A distant relative and friend, Gopi Ram, became my partner. Through Gopi Ram and a company owned by other relatives, we arranged and paid Rs 50 lakhs to Rajinder Mittal for Swamiji. In the bargain, I had to commit to sharing 35 per cent of my earnings with Gopi Ramji and this company. I could keep 65 per cent.

Normally, the annual bilateral agreement between India and the USSR was signed in December for the next calendar year and the deals were done in January-February. This time, the bilateral agreement between the two governments was signed in March. I had paid the money to Swamiji but there was no word about when the buyers' team from the USSR would be coming. I waited till the end of April but the team did not turn up. I was wondering what to do. Instead of waiting endlessly, I decided to go ahead with my original plan of visiting Europe to explore the possibilities of expanding our packaging business. The plan was to attend a few packaging exhibitions in Dusseldorf and Milan.

I was also eager to do some sightseeing in Europe. I assembled my brother Jawahar, Gopi Ramji and my friend Subhash Grover and told them that I was going to Europe. In the unlikely event the Soviet buyers came, they were to brave it out and sign the contract. In any case I would be back in about ten to twelve days. They were not happy or confident about this but put up a brave face.

I knew as well as them that it would be an impossible assignment for them to handle, but we thought that the buyers would not come to India till I returned.

I STARTED THE preparations to embark on my first-ever foreign journey. There was a lot of excitement in the household; wish-lists were coming in from family members about what they wanted from abroad. Those days a limited amount of foreign exchange was released irrespective of the duration of the trip. I had asked Purushottamji, landlord of the packaging factory, to travel with me, as he knew more about the packaging industry than I did. We had limited funds to spend and our travel agent told us that no medium-budget hotels were available for the trip. We decided to take membership of YMCA (Young Men's Christian Association) and stay at their hostels. Wherever they would not have hostels we would have to take our chances and scout for a budget hotel.

We wanted to carry extra foreign currency but that was not allowed. We had to resort to trickery. An old friend, Makhan Lalji, arranged for a suitcase that had a false bottom where we could hide two thousand dollars. In April 1981, Purushottam and I embarked on our journey. Many family members and friends came to see us off at Delhi airport.

The trip in itself turned out to be adventurous. We stayed at YMCA Dusseldorf and later travelled by train to Italy without realizing that we needed a visa not only for Italy but also for all countries the train would pass through. On the way back, while the train was passing through Switzerland, we were caught and were about to be offloaded. Luckily, an exhibitor from a packaging event recognized us and gave a personal assurance to the visa officer. The officer issued a transit visa to us on the spot and allowed us to continue on our journey.

Time flew when we were in Europe. We met several technology companies and assessed several new machines. Finally, we decided to start a project of laminated tubes. We signed a letter of intent at the exhibition itself with a technology-cum-equipment supplier from Switzerland named KMK Machinen AG. Their general manager, Bernhard Schwyn, became a close friend later and I was asked to be godfather to one of his newborn sons.

While we were roaming around in Brussels, I started feeling uneasy. We had been eating and shopping like tourists but I felt uncomfortable about something. This feeling was getting stronger by the minute. I told Purushottamji that there must be something wrong at home and that we should call and check. We had not been in touch with them for a week after coming to Europe. We stopped at a public call booth and connected to India.

Subhash Grover came on the line first but sounded very nervous. Jawahar and Laxmi almost jumped on me when they came on line. They had been trying to contact me for some days and did not know where I was.

They said that the Soviet trade commission had been calling repeatedly since the buyers had reached Delhi. And even Vijay Dhar had called several times. Dhar had scolded my brothers when he learnt that I was travelling.

How can you allow him to go at a time like this, Dhar had thundered.

I told my brothers on phone that it was not a big issue. The deal had been done, all they had to do was sign the contract. They could do it on my behalf.

My brothers had met the representatives of the Soviet delegation in Ashoka Hotel. But when they were asked rice-related questions, they were dumbstruck. They did not have any answers. The Soviets then demanded to speak to someone who knew the business. The Soviet buyers were wild at my team, 'You all do not know whether rice is grown on trees or above the ground or under the ground, how can you do business of 50,000 tons and how can I rely upon you?'

My brothers insisted that I return immediately. They made it clear that the deal would not happen without me. I decided to return.

9

OF RICE...AND AVARICE

*The Soviets smell something fishy
about basmati!*

I CALLED MY travel agent and took the first available flight back.
I was in Delhi the next morning.

Vijay Dhar scolded me sharply. But the deal was signed the
next day, on 21 May 1981. The deal was to export 30,000 tons of
rice to the USSR. I came to realize later that if I was not away we
would have got all 50,000 tons. The bad timing of my trip cost
us a loss of 20,000 tons of business. The buyer naturally wanted
to help his old friend Tulsi Tanna and gave him the business.
The Soviets told Swamiji that we were unreliable and we could
not be trusted with the entire consignment. It's remarkable how
a sense of unease had forced me to call home. If I hadn't called,
I might have missed out on the deal altogether as the Soviets
would not have waited for me.

There was a twist in the tale, though. Those days my official
name was Subhash Chandra Goenka. The Soviet official said,
'Mr Goenka, you don't know this trade, but we can help you.
We have given a contract of 20,000 tons to our friend and old
business associate from Mumbai. To you we are giving 30,000
tons because of the recommendation of our common friend.
You know which one, right?'

I nodded vigorously.

The Soviet asked me to meet Tulsi Tanna to take advice on
the deal. I went over and met Tanna, who till then had enjoyed
a monopoly over rice trade to the USSR.

Tanna welcomed us with sweets and made a strange offer.
He said he would fulfil our contractual obligation and make the
actual delivery of the rice for us. We did not have to worry about
the actual purchase and delivery of rice. He offered us a per ton
profit for doing nothing and promised to handle everything.
Basically, he was paying us a commission on our contract that
he wanted to handle directly.

'The rate for buying and processing one ton of rice is
Rs 5,000. But I will buy the rice and deliver it. You pay me
Rs 4,300, the rest is your profit,' Tanna offered.

I sat down with him and bargained over a few rounds of
whisky. After a while, he made a better offer. He asked me to
pay him Rs 4,000 per ton. My profit would be Rs 1,000 per ton.
Swami Dhirendra Brahmachari and Mittal were expecting
Rs 350 rupees per ton according to the profit-sharing ratio
agreed between us. I calculated a net profit of Rs 650 per ton for
ourselves.

For this profit, I did not have to do anything. The entire
transaction, from procurement to delivery, would be handled
by Tanna. In theory it looked like a good deal. But I didn't
commit to anything. I did not know much about the rice trade,
I felt like an *anari* or a novice. I needed some time to understand
the business. Jawahar and Gopi Ramji were with me and they
also thought that we had to discuss this matter before confirming.

I told Tanna that I would think about it and get back to him.

When we came out, Gopi Ramji remonstrated with me: 'Do
you have any idea how much you will make even with Rs 500
per ton? You will make about Rs 15 crores. And with Rs 650 per
ton, you will make Rs 19.50 crores.'

'You have rejected a smooth Rs 15-20 crores for no effort. I think you should return and finalize the deal right away,' he insisted.

But I was curious. I wanted to know how Tanna would make a profit if he received only Rs 4,000 per ton. If Tanna was happy with Rs 4,000, then clearly there was more to it than met the eye. I prevailed on Gopi Ram and we left.

Now we started our effort to understand the rice trading business. I spent the next three or four days in Naya Bazaar talking to traders. We figured out the different types of rice grains being traded and their price.

The basmati that is exported is not pure, we were told by multiple sources. It was mixed with parmal rice. Parmal is the same category of rice as basmati, as per the classification, except that the basmati grain has fragrance and is a weak grain. Therefore, it has a higher percentage of broken grains in it.

Then I did some calculations based on the market rates of basmati and parmal. I figured that we could procure and deliver the rice for about Rs 3,500 per ton. Now I could understand Tanna's motive. At this rate he would make Rs 500 per ton. He was making Rs 15 crores off us! And that too in the guise of helping us.

It was obvious that we should manage this deal ourselves, but Gopi Ram reminded me of the key obstacle. We still faced the shortage of working capital. Even though I had paid an advance of Rs 50 lakhs, I still needed Rs 2-3 crores to start work as a rice supplier. The family effort would raise about another crore for us. But buying 30,000 tons of rice and delivering it was a huge task. We would need at least Rs 3 crores almost immediately. I told Gopi Ram that we would figure out a solution.

I had exhausted all options but I was not willing to give up this opportunity. I did not want Tanna to handle this deal. I saw this as a chance to enter a new business on a scale that I had never managed before. It was a challenge that I wanted to take head on.

THE ONLY OPTION was to borrow from a bank. So I went to State Bank of India's branch at the Delhi Milk Scheme premises where we had our account. I showed the contract letter to the manager and said that I wanted to borrow for export. The manager did not believe me at first. He took out his glasses, polished them and pored over the document for a few minutes. Finally, he was convinced but said that lending for this business was done by SBI's overseas division. He offered to accompany me and introduce me to the relevant bank officials.

This was the first time that I had approached a bank for a such a large loan. We wanted Rs 6-8 crores of packing credit for the business.

The overseas division of the bank prepared a proposal of export credit for Rs 8 crores with a request to release Rs 2 crores immediately, and sent it to their head office.

Soon after, we started the process of purchasing the rice. We appointed a dealer to buy rice from various millers and began to aggregate it in our warehouse/factories. A lot of it was on credit from our dealer. We promised to pay him later. I tapped a friend from school, Suresh Jain, who was in the jute bags packing business. He gave us a source who supplied the required number of jute bags. He was of great help as he was patient about getting his payment. With such friends I could get more work done with less money.

The Rs 1 crore I had raised from the family disappeared in payments for truck freight and labour. Even before I had begun the export, I owed about Rs 15 crores to traders and packers. We had begun collecting the rice and started despatching by rail and road transport to Kandla port. Other procedures included quality inspection by the buyer's surveyors and some formalities specified by the government agencies. The rice was to be loaded onto Soviet flag vessels at Kandla port.

I still had not got the first Rs 2 crores credit from SBI. I couldn't bear the delay any longer and went to the manager (exports) and had a long conversation with him. I explained the challenges and problems and pleaded for his help. He was moved by my emotional talk and escorted me to the chief manager's office at the State Bank overseas branch at Parliament Street in New Delhi, and requested his intervention.

I told him that we had sent goods worth Rs 15 crores under the lien of the bank. I said our credibility in the market was at stake. We had issued cheques to our rice dealer for about Rs 2 crores. The cheques were presented but could not be cleared without the sanction from their head office. I told the chief manager that it was unfair that the bank was benefitting but not releasing my loan.

I threatened him that I would take my business to another bank if he did not help in releasing the money that very day by clearing the cheques. It was an empty threat, since I had no backup plan.

Luckily for me, it worked. The chief manager called the Mumbai head office and forced them to take a decision. The sanction was received on the phone and the cheques were honoured. Everyone had been on edge. It was only after the release of the loan that the tension was relieved.

Meanwhile, another crisis had cropped up. Since I had taken a lot of rice on credit, there was a buzz in the market that I did not have money anymore and that the deal was going to fall through. Traders gossiped that rice was not being loaded onto the ships and that we were about to go under. This was bad news for me; it could have stalled my procurement of rice. I had bought about 10,000 tons, and more was in the pipeline.

Now I approached the agent of the ship that was to carry the cargo to the USSR. The agent company's owner was one Naresh Kotak. He was based in Mumbai but was in Delhi for a visit and was staying at the Oberoi hotel. My shipping agent had advised that if he could get a nod from Kotak he would issue the bill of lading (a document issued by a carrier detailing the shipment). I would then get paid for the rice awaiting shipment at the port. This would stop the rumours and also the creditors would be paid and satisfied.

I did not know Kotak at all. With great trepidation I called the hotel and asked for him. I spoke to him and asked for a meeting. He had heard my name and the fact that I had won the prized deal with the USSR. He seemed to know that I had got the contract because of my connections. He immediately invited me to the hotel. My anxiety in approaching him was not justified after all. My reputation had preceded me. I did not know how many ripples this deal had caused. An unknown businessman like me had got a contract that required a lot of influence.

When I reached, he offered me a nice whisky and we chatted. I told him that we had despatched the consignment of rice. His company had received about 10,000 tons of rice at Kandla and now I needed the bill of lading (a document issued by a carrier detailing the shipment) for it. He told me that it could be issued once the cargo was loaded on the ship. I reminded him that he was the agent of shipowners as well as our stevedoring agent. The large quantity of rice of approved quality was in his possession, and he had the power to issue the bill. After some convincing on my part, he agreed to issue it.

I told him a bill of lading for even 4,000 tons would help. He called his office in Bombay and spoke in Gujarati, which I did not follow much. He then promised that the bill of lading for 4,000 tons would be given the next day. He advised me to send someone to collect it from Bombay. I was grateful and nervous at the same time. I thanked him and left.

I couldn't sleep that night. This bill of lading was critical for the success of the deal. Instead of sending someone, I took the 6 a.m. flight to Bombay to get the document personally and returned the same evening. That night also I had very little sleep as the next day we were to negotiate the letter of credit with the bank.

Now I had to put all the papers together to fulfil all conditions laid in the letter of credit. We had to fulfil ninety-four conditions in the form before the money could be released. None of us in our office had any knowledge of export documentation. I asked my friend Grover, who had worked in Nafed. He did not know either. I got so exasperated that I called my secretary, a south Indian girl, and began working on the papers myself. It took me about 3 hours to get everything done.

I reached the bank with the documents. The official spent a couple of hours going through them. He found three or four errors in it that I could correct on the spot, but the rest was fine. Now we had another Rs 2 crores. Meanwhile, the SBI head office approved the remaining Rs 6 crores. Now we were through.

Meanwhile, the ship for our rice consignment, which had been waiting for three weeks, finally got the go-ahead to pick up the cargo. The first consignment sailed from Kandla port to the USSR. By now about 15,000 tons of rice was in various stages of procurement, processing and shipment.

AFTER OUR FIRST consignment went off well, we were relieved and delighted. I had taken rice worth crores of rupees on credit, managed shipping issues and export documents—despite having no experience in this work. That was what excited and challenged me. I was ready to take on more complex orders from the USSR. But before that, I had to ensure that my consignment had arrived safely in Moscow and had been approved there.

Tanna Exports was not happy that we were succeeding. Not only were they spreading rumours in the Indian market but were telling officials in Moscow that the rice supplied by us was of inferior quality. A few days after the ship left, I took an Air India flight to Moscow to confirm the delivery and to make sure that the buyers were happy. But there was more trouble waiting for me when I reached the offices of the Exportkhleb.

The moment I entered their office, about five officials cornered me. They appeared angry and agitated. One of them said, 'We are very upset, Mr Goenka. We have received your first shipment and are disappointed. You have sent us inferior-quality rice. This is "B-Grade" rice. Our other supplier (Tanna) has dispatched A-grade rice.'

I was shocked. I couldn't believe what I heard. I had been very careful in processing the rice as per the advice I got from the traders. I had asked a rice expert to oversee the entire process. He was based at Kandla port to make sure the quality was not compromised. I had ensured that at least 50 per cent of the rice was basmati. The basmati was the finest grade, while the other 50 per cent was parmal rice. Parmal was the same grade of rice as the basmati rice. I had been told that it was the prevalent business practice.

For a few seconds I couldn't think of what to say. But I managed to ask, 'May I see a sample of the A-Grade rice that you have received from Tanna?' I wanted to compare my rice with the competitor's consignment.

Shortly, another officer brought two plastic pouches of rice. One was mine, the other was Tanna's. They opened my sample first. Now, basmati is a thin grain and breaks easily. When it is being de-husked, the pressure applied to it is lesser than for normal rice. But the result is that each grain has a thin red line that marks the place where the husk was. This red line can be easily seen, but disappears when it is cooked.

The Soviets saw that some of the grains had these red lines. They thought this was a sign of low quality. And therefore called it B-grade.

When I compared it with Tanna's sample, I understood the problem. Tanna's sample didn't have a single grain of basmati! It was all parmal rice. The poor Soviets had no idea about the difference between basmati and non-basmati rice! And the wily Tanna had been sending them 100 per cent parmal rice, which was obviously cheaper. Basmati cost Rs 3,000 per ton while parmal was half that at Rs 1,500 per ton. He was passing off cheaper rice as basmati and could get away with it since he had a monopoly on exports till then. Tanna was making a killing even at the low rate he had offered me.

It would have been impossible for me to educate them about the difference between basmati and parmal rice. According to Government of India's classification, both types of rice were of superfine variety.

This was the first time I was trading with them and was in the process of establishing our credibility with the Soviets. If I tried to prove that Tanna had fooled them, it would reflect badly on them too.

I think Europeans have never enjoyed the aroma of hot, cooked basmati. They did not have any concept or appreciation of fragrant rice, at least till then.

So I did the only thing I could. I apologized.

'It's my mistake, I will correct it immediately. The next shipment will be absolutely A-Grade and according to your expectations,' I assured them. They were mollified and I left right away.

I called Delhi immediately and asked for all supplies to be stopped instantly. Even the consignment that had reached the port was to be returned.

Gopi Ram was aghast. 'Did you have too much vodka with the Russians?' he shouted on the phone.

I replied, 'These Russians took my pants off. The rice we sent was treated as B-Grade. They would have rejected our entire shipment, but they did not because of our friends who had recommended us. They want only parmal rice. Please make sure that only parmal rice is loaded on the ship now. All the basmati rice we have procured should be returned or sold back.'

So our entire team worked to ensure this. Our future consignments did not face any further complaints. This was a rare case of a lower-grade product finding greater acceptance than the superior one. Such was my unique experience of exporting rice to the USSR!

10

A LATE-NIGHT MEETING
WITH MRS G

*Feeling like an ant in a fight
between elephants*

DHIRENDRA BRAHMACHARI SEEMED to be focused on money and was always open to new opportunities. One Bhaskar Bhattacharya used to visit him from the US. He used to hold long discussions with Swamiji to convince him of the need to convert the public service television broadcaster Doordarshan (DD) into a commercial service.

I had met Bhattacharya, too, but whatever he said about commercial TV went above my head. I had no understanding or interest in TV or the broadcasting sector at the time. Of course, I had no idea then that destiny would one day take me to that very world.

Dhirendra Brahmachari did his bit for the commercialization of Doordarshan in 1983. He hosted a yoga show on DD. He could be convinced to experiment with new ideas if there was something in it for him.

I used to meet him twice or thrice a week just to keep the relationship going. He was usually surrounded by five to six young women. I was not surprised because he had a magnetic personality. Women found him very attractive. He would wear a dhoti all through the year, even in the cold winters.

While he helped me with the rice deal, it would also lead to his downfall. He fell in the esteem of the Gandhi family because of his role in the rice trade with the USSR. Apart from me, barely two or three people knew the real reason for Brahmachari's fall from grace with the Gandhis.

It happened thus: The contract for 1983 was to be decided in December 1982. Now Brahmachari sent instructions that all the future export of rice would be done by his own newly formed company. He also told Dhar about this.

I was told that I would not get the contract any longer. This was despite the fact that he had taken an advance of Rs 2 crores from me towards profits for next year's contract. I said that it was fine if I did not get the contract, but that the advance be returned to me. He refused to give it back. He thought I had earned more than I deserved and hence did not feel the need to return my money.

I had no choice but to keep quiet. But one day Vijay Dhar called me for a meeting, and asked, 'What has happened between you and Brahmachariji?' I could figure out that he had heard about Swamiji's decision to start rice export himself, or perhaps through Mittal and or some other traders.

Those days there were two groups close to the Gandhi family. As normally happens with any power centre, satellite power centres emerge around the main players. There was a group known as 'the Kashmiri group', which included M.L. Fotedar, Arun Nehru and Vijay Dhar. The other group included R.K. Dhawan, Dhirendra Brahmachari and some others. Both groups were rivals and wanted to corner power. I told Dhar about Swamiji's decision to export rice through his own company. And also that he had decided not to return my advance for the year's export order. 'How much money have you paid him so far?' Dhar asked. I sincerely revealed the number to Dhar as he was key link between Rajiv and me.

Dhar asked me to wait in his office and went to Rajiv's room next door. Rajivji came back with him and began to ask me a lot of questions. I had to reply to all of them. Somewhere at the back of my mind I knew that I was getting into a bigger mess. And that too for the loss of a few crores' advance. Rajivji was very surprised to hear what I had to say. It appeared to me that Swamiji had given them a wrong picture about our profit-sharing deal.

But now I was worried, 'I don't want to get into a fight, sir.' I told Rajivji. 'You are powerful people and in such a conflict between big personalities, a person of my stature will get crushed as if I never existed. If you can help me get my money back I would be grateful. But if you can't, that too is fine. I will assume that this was not in my destiny.'

Rajivji took my plea seriously. He assured me about my future. 'But you have to come and say everything to someone I have in mind,' he said.

I guessed that it had to be Mrs Gandhi. Who else could it be? The prospect of facing Indira Gandhi terrified me even more. 'Please don't involve me in this,' I pleaded. I had never really met Indira Gandhi on a one-on-one basis. I had just seen her at their house when she was out of power. She wouldn't even recognize me, I thought. I kept quiet and the meeting ended.

A few days later I was asked to reach Prime Minister Indira Gandhi's residence at Safdarjung Road at 9 p.m. Before leaving home, I told my family that I might not return at all that night. That's how terrified I was. I had no idea what would happen next. I asked Jawahar to book me on a London flight the same night. I told him that I might have to leave India that night if there was trouble.

I was a thirty-two-year-old trader from a small town. And I

was at the centre of a fight between two powerful groups around the ruling family of the country. I wasn't close enough to any group to receive protection. I said to myself, 'these elephants are fighting and you will be crushed in their tussle'.

On the appointed day, I reached the prime minister's residence by 9 p.m. and was asked to wait. Indira Gandhi was supposed to fly to Europe later that evening. I waited for half an hour, then another hour passed. Two hours passed. The wait was even more terrifying. It was a December night and fairly cold. Finally, I was summoned inside the room around 11.15 p.m.

Sitting in the room were Indira Gandhi, Rajiv and Dhirendra Brahmachari. This was 1982. Rajiv was not in the government but was general secretary of the ruling Congress party.

There was silence for a few seconds. They looked at me very closely. I almost peed in my pants.

Mrs Gandhi looked at me closely and spoke first. 'I thought you were an older person. But you are very young,' she broke the ice.

'How much money have you paid Swamiji?'

'About Rs 2 crore advance for this year,' I said.

'No, no. I want to know how much have you paid in total,' she persisted.

I told her the figure.

Swamiji's eyes were blazing. He was looking at me with deep hatred and anger. I could see him from the corner of my eye. I felt like a mouse surrounded by hungry cats.

I was asked two or three questions and then I was allowed to leave the room. As I was leaving, Rajivji asked me to wait in the other room.

He came out after an hour and said, 'Congratulations, now go and relax.' That's all.

I left right away, relieved but confused. It was 1 a.m. or so. I had spent about four hours in absolute terror. I reached my Punjabi Bagh home by 1.30 a.m. and hit the bottle. I was too nervous to sleep.

From that day onwards, Swamiji's downfall began. I think the Gandhi family did not trust him completely after that.

We made huge profits of many crores on this trade, which I conducted from 1981 to 1984. This came to an end around 1984 when I was told that there was a lot of pressure from various quarters on Rajiv Gandhi to recommend some other person for such contracts. People felt I had benefited enough and that others should also get a chance to earn. Vijay Dhar called me and said that there was too much pressure to give this contract to a Delhi-based businessman close to the Congress party. I told Dhar, 'Bhai Sahib, it is fine by me, I have made enough money.' I thanked him profusely. I did not try to hold on to the contract or try to persuade Dhar otherwise. I must concede that the Gandhis were magnanimous and still recommended me for one half of the contract. The other half went to the businessman from Delhi. That was the last year I got the rice contract. From 1985 onwards, the Delhi businessman got the entire contract through his Congress party connections. Tanna had continued his work independently since he had direct links with senior officials of Exportkhleb. He was not dependent on the recommendations from India.

Families such as those of the Delhi businessman and Amitabh Bachchan were close to the Gandhis. Compared to them, I barely knew them. I think they were good to me because of my small but insignificant help when they were out of power. The Gandhi family had an important trait. They never forgot people who helped them, especially when they were not in power. They remembered the small help I had extended during their difficult days. Rajiv knew about my assistance to Sanjay through Tripathi. This was also an important lesson for me. I learnt to ensure that I kept my relationship with people irrespective of their position.

The contract stayed with us for four years. The first two years we shared the profits with Dhirendra Brahmachari and the next two years with Sitaram Kesri, who was a treasurer of the Congress party. The last couple of times I was asked to deliver the money at Arun Nehru's house. But I never met Arun Nehru; Vijay Dhar remained my key contact.

Many years later, after Rajiv was assassinated, I was told that the Gandhi family believed that much of the profits from the rice trade were siphoned away by Arun Nehru.

I had no one I could explain that I followed the instructions of Rajiv's pointman. I didn't even know how I should clarify and to whom. This remains an issue with me till the writing of this book. I wish I could clarify my position to someone. They did not know that I met Arun Nehru only in 1993 when Zee TV gave him an award for being the best political commentator.

Before the Delhi businessman got the contract, several other people tried to get it. Many MPs wanted the contract. In fact, some stories appeared against me in the newspapers. In December 1981, *The Statesman* had a front-page story about how a young businessman from Haryana was being helped by the Soviets. The report said that my profit was equal to the budget of Haryana state. Nobody knew that Rajiv was supporting me. They assumed it could be someone within the system.

I managed to keep the contract partly because nobody knew about Rajivji's support for me. For the first few years, even my brothers did not know that Rajiv was my supporter. They thought that Swami Dhirendra Brahmachari was backing me. I did not mention Rajiv's name to anyone. In fact, I did not mention Rajiv's name to anyone till 1999-2000, many years after he had passed away.

During the period I was doing business with the USSR, I also became a messenger for requests for meetings between Rajiv Gandhi and the Soviets. I would call Vincent George or the other secretary, Madhavan, to organize the meeting. The code name we had for Rajiv with the Soviets was White Trousers. The Soviet trade representative would pass messages to meet him through me. They wanted to maintain a distance in public. Rajiv and the Soviets trusted me because I was discreet.

AFTER THAT LATE-NIGHT meeting at Indira Gandhi's residence, I came a bit closer to the family. I could visit them if I wanted.

One day Dhar called me with another offer. He wanted to make me an agent for a bigger deal. I asked him what it was. He said it was to import arms for the defence services. I considered it for a while but declined. I couldn't sleep for three days after this offer. I did not want to be a *maut ka saudagar* (merchant of death). Eventually, Win Chadha was appointed an agent for the deal. The rest, of course, is known to everyone.

In the meanwhile, we had started exporting other products to the Soviets, too. Since we had an equation with them, we exported peanuts, oil extractions, etc. At our peak in 1983, our exports to the USSR were about Rs 150-200 crores. Our rice export contract went up to 1.5 lakh tons in one year. It was not a smooth ride—we faced a few hiccups. The income tax department slapped a notice on us for evading tax. They calculated that we had made a profit of more than Rs 200 crores. After all the scrutiny, they accepted our declaration of profits.

We expanded the range of products for exports and the countries we sold to. We exported carpets, textiles, shirts and other such merchandise. We created another company called Essel International for this.

Dhirendra Brahmachari's situation worsened after Indira

Gandhi's death. Despite the money he made, he had fallen on hard times in the last years of his life. I gave him Rs 50,000 a month as support from 1993 till his death in 1995. It was out of a sense of gratitude. He was instrumental in my success in the rice trade. I had made crores of rupees with his support. It was the least I could do when he was down.

11

GOOD THINGS COME IN SMALL PACKAGES

After teething troubles, we hit pay dirt

WHILE WE WERE exporting rice, the work on the packaging business had increased as well. As mentioned earlier, during my visit to Europe in 1981, we had decided to bring laminated plastic tube technology to India. We had signed up with the Swiss company KMK Machinen AG, and the American Can Company. Both of them had joined hands to supply the equipment and license the technology.

Traditionally, tubes used to be made of aluminium, with the head of the tube made of zinc and lead. Laminated plastic tubes were developed in the US after a controversy. Sometime in 1974-75, a family in America had sued Proctor & Gamble for using lead in toothpaste tubes. A child had chewed the toothpaste tube, and the lead on the shoulder of the tube had caused ulcers in the child's stomach. Since P&G had not put any warning on the tube, it was sued. The company then decided to look for alternatives to the metal tube.

The American Can Company of the US developed a tube where the metal sheet was sandwiched between two layers of plastic. The Swiss company KMK also helped develop this technology. And then companies like P&G began using it globally. They added layers with more plastic, and even paper. This made the tube less flexible and also protected the content from external environment. But this technology had still not come to India. Both the American Can Company and KMK together were licensing the technology, and we signed two letters of intent—one for technical collaboration and the other for buying the machines.

We bought the technology not just to make the tubes but also the raw material for making them. We had signed the letter of intent in 1981, but then got busy with the rice exports. After the first shipment of rice had gone, I started work on the packaging idea. I commissioned a market study by Tata Economic Consultancy Services (TECS) on the need for such tubes in India. The study said it was a great idea, but was futuristic. The investment would be high and therefore, the breakeven percentage would be higher. The cost of such tubes was also high and therefore, consumer products companies might not go in for the tubes. TECS suggested that these tubes should be introduced in the Indian market after six to seven years. They argued that the cost of technology and equipment would fall after a few years, making the project in India more viable.

But I was set on making these tubes and decided to go ahead with the project right away. Eventually, though, the TECS study was proved right. The final return on investment took longer than I thought.

I bought the tube-making machine from the Swiss company. Its capacity was about 60 million tubes per year. The printer capacity was also the same. But the lamination machine I bought from Sumitomo of Japan had a higher capacity. It could make sheets to produce 600 million tubes. This was a total mismatch, but I could not do much about it. A smaller machine was not available. The entire investment on setting up the factory was about Rs 7 crores. But now I ran into another set of issues.

In 1982-83, any industrial activity that needed capital of more than Rs 5 crores needed a licence from the industry ministry, as it was classified as a large-scale undertaking. This was a very lengthy and complicated process. India was still in the grip of the licence raj. For a large unit, apart from the Central government licence, a no-objection certificate from the state government was required, in addition to more than two dozen other clearances.

We had thought of setting up the factory in Dharuhera, near Delhi, in Haryana. We applied and got an NOC from Haryana. Unfortunately, we had applied for this even before I got the TECS report, which said that all buyers of tubes were based in western India. Thus, making tubes anywhere else would make it unviable. The cost of transporting the empty tubes would be prohibitive.

Now I needed an NOC from Haryana allowing us to shift the project to another state. But Haryana did not want us to shift the project. I had to use all my contacts and clout in Haryana to get this done. Cancelling the first NOC was tougher than getting it!

Now I had another hurdle to cross. I had to get the licence from the industry ministry amended since I wanted to change the location to Maharashtra. For this I made many efforts but our request was rejected twice. Finally, I had to approach Vijay Dhar for help. He called the secretary in the ministry of industry and asked him to meet me and help. I reached the secretary's room at Udyog Bhawan. He asked many penetrating questions. Why do you want to leave Haryana? Why Maharashtra? Why do you want to change? Changing location of factory is against our policy... Such was the atmosphere at the time. N.D Tiwari was the Union industry minister then.

Those days, an industrialist did not have the freedom to change the location of his plant. Even to set up a plant, one needed to convince the government about its economic viability. This, when we had not taken a single rupee or investment from the government. Finally, since the recommendation was from Rajiv Gandhi's office, the secretary, industries, allowed the location to be changed from Haryana to Maharashtra.

I knew that the plant would have to be near Bombay. But I could not decide on the exact location. I happened to meet a friend, Prithvi Jindal, on a flight. His family owned and operated pipe and steel factories at many locations in the country. I asked him for advice. He had set up his own large-scale factory in an area just outside Mumbai. He recommended the same area. I felt more assured since I would have a friend's factory in the neighbourhood.

WE REGISTERED THE company as Essel Packaging Ltd., and created a trademark of Lamitube for the tubes. While we started work on acquiring land and setting up the factory, the next step was to get our buyers in place. I approached all the toothpaste manufacturers in the country, including Colgate, Ciba Geigy and Hindustan Lever.

Hindustan Lever was one of largest producers of toothpastes and therefore a large buyer of tubes. Colgate had the largest market share in toothpaste. When I met the head of manufacturing of Colgate, he refused to accept our plastic laminated tubes, giving several reasons. He could not produce more than the licensed quantity allowed by the Government of India. He had more than one dozen suppliers of aluminium tubes. And he had no reason to switch to a monopoly supplier. If he shifted to my laminated tubes he would have to make extra investment in setting up new filling lines for the toothpaste.

This head of manufacturing was probably very close to the aluminium tube suppliers and therefore was not willing to move to a more progressive packaging material. I realized that other toothpaste makers would give me the same reasons for not buying my tubes.

I gave this feedback to our technical collaborators. They, especially the Americans, were upset. Since Colgate was an American company, they talked to its headquarters in New York. They hoped to get directives issued to Colgate India to switch from metal tubes to laminated tubes. We were happy to hear that, and I took a trip to the US. We met senior executives at Colgate headquarters along with the vice president of American Can Company, but nothing happened.

We were getting desperate. I made a strategy to address all concerns of the Indian toothpaste makers. We agreed to reduce our price per tube to a level that barely covered the cost of manufacturing. I offered to bear part of the capital expenses for setting up the filling lines in the factories of the toothpaste makers. Their concern over my being a monopoly supplier was natural. Our monopoly would be broken only if another company set up a similar manufacturing plant. I did not want an aggressive company to become our competition. I took a strategic decision to get a friendly company to set up another plant. I asked my old friend Anil Chanana to invest in a plant with our technology support so that we were not seen as monopoly suppliers. Despite the rice export discord, Chanana and I had remained friends. By ending the deal with him, I had preserved my relationship.

Our efforts to talk to toothpaste makers seemed to pay off after a while. Our meeting with Hindustan Lever moved forward and they appeared ready to buy our tubes. A joint task force was set up to work out the details, including making changes to their filling lines for the toothpaste.

Finally, when they agreed to buy our tubes and booked our

entire capacity of 60 million tubes per year, we were pleasantly surprised. They signed a letter of agreement. Shunu Sen, the sophisticated, soft-spoken head of personal products division, himself was the chief guest at the inauguration of the factory in 1983.

After the inauguration in 1983, we raised money from the market. I launched an IPO for Essel and it was oversubscribed more than forty-one times.

But there was a shock in store for us. Despite the agreement, it appeared that HLL had no intention of buying the tubes from us. They were not interested in changing their tubes from metal to laminate. They realized that laminated tubes were the future but also wanted to prevent their competition from using our tubes.

When we were ready to supply, they kept postponing the delivery. They kept us hanging for almost two years like this.

We were inexperienced and Levers took advantage of that. The contract we signed had placed all the duties on us, but no rights in our favour. It was a one-sided contract. We could not recover anything from them for not fulfilling their obligation to take delivery of the tubes. They had cleverly booked all the capacity with us and stopped us from marketing or selling to any of their competitors.

There were enough reasons in my mind to believe that HLL in India was a terrible corporate citizen. Our contract bound us to supply all tubes to Levers. It did not allow us to sell to anyone else. Worse, it did not mention any deadline for payment or acceptance of delivery.

It was now 1985 and not a single tube had been sold. In our rush to get HLL as a customer, we had signed the agreement with them almost blindly. Now we were suffering for that huge mistake.

The costs were mounting and the Rs 3.6-crore capital of the

company was almost wiped out. There was no manufacturing as there were no buyers. But I had engineers, production controllers, and labour on my rolls. There was interest to be paid on the loans we had taken.

We survived because we were still earning from rice exports. But we were close to becoming a sick unit. The rest of the group provided interest-free capital to the tubes factory so that it could avoid the provisions of becoming a sick undertaking.

WE GOT SOME good news soon. There was a new player in the toothpaste market. The Parle Group launched their factory to make toothpaste under the brand name 'Prudent' in Bahadurgarh in Haryana.

After waiting for two years, I had stopped caring about my agreement with Levers. I was so desperate that I sold the tubes to Parle by breaking my HLL agreement. HLL had no right to stop me, if it was not buying from me. The first Prudent toothpaste that hit the market was in our tubes. The toothpaste was a market success as Prudent used the easy squeeze feature of the tubes as a terrific marketing plank. The consumers loved the convenience of the tube. Prudent had a great launch, thanks to the smart decision to use modern packaging.

This really shook up all the players. Suddenly, HLL realized that they were on a weak footing. The success of Prudent forced even Colgate and Godfrey Philip to come to us for tubes. P&G, the original user of such tubes in the US, did not make toothpastes in India, and therefore did not approach us.

Soon, the export division of HLL contacted us and offered to buy our tubes. This unit was different from the domestic sales unit. They made the changes to their lines for our tubes. Then Ciba Geigy came to us for Binaca toothpaste. Lever's domestic division finally decided to take our tubes in 1986. By then our factory was running to full capacity.

After that, there was no looking back. We set up four more factories. A point was reached where some buyers complained that we were monopoly suppliers.

After years of struggle, our commitment to laminated tubes had paid off well. Essel Packaging became so successful that I bought the Swiss company that made the tube-making machines. Now we are the biggest tube makers in the world— manufacturing about 8 billion tubes a year. One third of humanity uses our tubes while brushing their teeth. Our factories are in all five continents.

By 1987-88 this business was in full swing. Once the business settled, I started feeling restless and felt as if I was unemployed. Ashok had taken charge of the packaging company and was running it brilliantly.

12

A ROUGH ROLLER-COASTER RIDE

Some people in government are
not amused

I HAD NOT yet started living in Mumbai, though I would visit the city on work or pass through it on my way to Kandla port to supervise the exports of rice and other commodities. In 1981-82, during one of these visits, I bought a piece of land in an auction by the Bombay High Court. A relative had told me that some land was being auctioned in Mumbai. Nothing was known except that it was within city limits. I didn't have the faintest idea about the geography of Mumbai. But I decided to participate in the auction. Any land within city limits had its uses, I reasoned.

Surprisingly, I won the bid. I had bid Rs 27 lakhs for the 753-acre plot. There were about six to seven other bidders but everybody had put in low bids. I did not notice any big builder.

I discovered the reason for the lack of interest only after winning the auction. The land was in a no-development zone. There were some ongoing disputes as well. Therefore, local builders were not interested in it. After buying the piece of land, I almost forgot about it.

Five years later, in 1987-88, when I was thinking of new

businesses, I remembered the land that we had. I decided to study what could be done on it, given the restrictions of it being in a no-development zone.

I found that most types of construction were not allowed in the area. Then I realized that the rules did not prevent the setting up of an amusement park. After discussions within the family we decided to set up such a park. During my travels I had visited several amusement parks, including Disney World and Six Flags, and had really enjoyed myself.

As this idea grew, I started looking for partners. Appu Ghar had started in Delhi in 1982 around the time of the Asian Games, but I wanted to set up a bigger park. I met several companies in Europe that owned and managed amusement parks. After many visits, we zeroed in on 'Blackpool Pleasure Beach' in the UK. The CEO of the Blackpool company was a former officer of British Army. When I suggested a partnership, he was a bit apprehensive. He thought he would be lynched if he returned to India! He had served in the army in India and thought everyone here hated the British.

I convinced him and his team to help us set up the park. Of the total area of 753 acres, we could develop only 80 acres. The Blackpool team designed the plan for the park. They sold us some rides, and we developed some on our own. We invested about Rs 30 crores in developing the park. But it was not easy.

The local authorities and the state government were looking at the project with suspicion. They thought that our claim of setting up an amusement park was an excuse, that our real motive was to use the land for real estate development, which was not allowed. They did not issue approvals for the park for many months. The government departments did not have any idea about amusement parks. They did not have any rules and regulations to go by. The local MP was Ram Naik of the BJP and he, too, was opposed to the park. This piece of land was on the coast and Ram Naik said there were fears it would affect the ecology of the area. We tried to convince everyone that the park would in fact enhance the environment.

We were looking for ways to convince the government about our intentions. Our PR agency had a good idea. The Central government was planning to celebrate the birth centenary of Pandit Jawaharlal Nehru in a big way. There was a Cabinet committee set up to plan multiple projects during the centenary year. The PR agency suggested that we should dedicate this amusement park to children. We would call it Children's Park in honour of Pandit Nehru's love for children. And if the Cabinet committee approved it, we thought that the state would give us its full support to develop it.

I went to Delhi and met P. (Bob) Murari, who had just stepped down as Cabinet secretary and who was coordinating/ heading the Cabinet committee for Nehru's centenary celebrations. He liked the idea and it was discussed in the committee, too. Rajiv Gandhi was the chair of the committee as prime minister.

I must say here that I did not have one-on-one meetings with Rajiv after he became prime minister. I met him with others but our interaction was limited.

When the project was suggested to the Maharashtra government, it received a lukewarm response. It was 1988, and Sharad Pawar was the chief minister. As it happened, the relationship between Sharad Pawar and Rajiv Gandhi was very prickly. Rajiv was not happy with Pawar and did not want him to continue as chief minister. Therefore, Pawar was not keen to help in any project suggested by the Centre. Thus the plan to include the park in the centenary celebrations fell flat.

I was back to dealing with the state government directly. I had to somehow convince Pawar. I used some common friends to request him to have the project cleared. He agreed to approve the project, but did not go out of his way to support it.

I was firm on my decision to build the amusement park. In order to expedite the project, I decided to stay put at the site. I bought a mobile caravan from the UK and stationed it at the park site. I started living in the mobile van to oversee the work. It was my office and residence for about eighteen months during 1988-89. Since permanent structures were not allowed in the area, we had to make do with temporary structures for workers and other executives. This was not new to me; I had stayed in a jungle-like area for a year in 1983 when the first Essel packaging factory was being built.

We slogged and struggled and managed to complete the first phase of the amusement park in eleven months. We had decided to name the park Essel World. Though the name sounds as if it was the natural extension of the group's name, this was not so. We had started a public competition to name the place. We had invited Mumbai's citizens to suggest names. We explained that the park would be built in five phases, and each phase was explained to the people. Out of the thousands of entries we received, they picked the name ESSEL (Educational Sports and Science Exhibition Land). Hence it came to be called Essel World.

Chief Minister Sharad Pawar inaugurated the park in December 1989. Even though Ram Naik attended the inauguration, he was vocal about his apprehensions. He continued his criticism even while he addressed the gathering from the dais.

Today, even after twenty-six years, we have still not received a municipal water connection to the park. Perhaps the municipal corporation believes its duty is to provide water to residents but not to tourists. This even though we pay a large amount of taxes and most of the visitors are from Mumbai. Initially, the city bus service run by BEST, too, refused to offer a connection to the park.

The worst was the taxation on the park. The government applied a 150 per cent entertainment tax on the entry ticket, as applicable to cinema at that time. While the price of a single-entry ticket was Rs 50, the tax was Rs 75. I appealed to Sharad Pawar again, pointing out that the park was meant for middle-class consumers. He instructed the chief secretary in my presence to 'look into the issue'. Though later, apparently, Pawar privately told him not to reduce the tax. But we continued to fight for lower taxes.

Four-and-a-half years after the launch, the park was to be developed further, in five phases. After the first phase, however, we could not develop it further because of environmental restrictions.

The troubles continued. The government revoked the entire sale of land that had been awarded by the Bombay High Court. We had to take them to court over it and the battle went on for years. In 2014, the Supreme Court held that the acquisition was legal and now the title of the land is with us. It took us twenty-five years to win the series of cases.

The power connection was given, but its quality was poor. We had to use our generator sets. Intially, in the absence of a telephone connection, we had to use the long-distance wireless network to be in touch with the city office and residence at Cuffe Parade. We got the phone connection only after six months. Still, amongst the government facilities, the phones were the most efficient.

Despite the high ticket price and infrastructural issues, there was a good response to the park. The first month saw about 180,000 visitors; in the second month it was 170,000. Slowly, however, the number started falling. Visitors found it tough to reach Essel World. The road connectivity was poor. The boats used by visitors to cross the creek and reach the park were smelly and poorly maintained. The monthly figure settled at about 100,000. The ticket cost Rs 125 per person. Despite poor connectivity, these people came as they were hungry for a new avenue of entertainment. Somehow, our governments never give any importance to such needs of the society. Any activity that generates tax revenues and meets the needs of society must be supported by the government. We were offering clean entertainment to lakhs of families but the government treated us as a toxic-waste producer. In the first year, we did not make losses, but there was no profit. Our market survey had estimated three-and-a-half million visitors in the first year. Only 40 per cent of this target was achieved.

We did another survey to find out what was holding back the visitors. The visitors said they found it tough to travel two to three hours to reach the park. Even though the park had enough activity for them to pass six to seven hours, they did not want to travel for so long on bad roads to reach there. They wanted the same entertainment to be available closer to their homes. They enjoyed the park experience but wanted entertainment with the comfort of better infrastructure.

This feedback sparked a new chain of thought in me. I realized how starved of entertainment Indians were. The middle class was growing and had started spending money on luxuries. They were no longer worried about basic needs. The government machinery was still geared for an atmosphere of scarcity and poverty. This machinery did not care much for a rising consumer class. To me this seemed like an opportunity.

Three thoughts swirled in my mind. One was how to offer entertainment to people closer to their homes. The second was an urge to expose the lethargy and inefficiency of government departments. The third was to make Essel World a destination with hotel facilities. All three of these ideas were a result of the pain I faced in launching Essel World.

13

BROADCASTING MY INTENTIONS...

...but no one takes me seriously

ALL THESE IDEAS bothered me for many weeks (even today Essel World is not the destination I would like it to be). I was looking for a way to meet the entertainment needs of people. Then I heard of one Dr J.K. Jain of Delhi. A BJP member, he called his company Jain TV, though he had no TV channel then. But he had started something very smart. He had built about eight to ten vans that had a video projector and screen. He would rent them out to political parties for their election campaign. The parties would take them to different constituencies and play their campaign films. I read about this in the papers. This was 1990.

I asked my team how many districts there were in the country. They came back in a few hours and said about 600. I thought I would place one video van in each district across the country. And that at one point during the day, at 6 or 7 p.m., we would show the same entertainment show on all 600 screens for the local viewers.

We thought it would be a great idea to embed ads in the shows to earn our revenues. We would get our money from the ads while the viewer would not pay to watch the show. The van would play the show in one town on one day and then move to another location. After every two weeks, we would change the programmes.

But before acting on this idea, I appointed a couple of executives to do a detailed study of the rules, regulations and taxes related to entertainment in each state. I had been so badly burnt by the Essel World launch that I did not want to take any chances.

When I got the report I realized it was going to be close to impossible. Each state had a maze of regulations on entertainment. Even if we were not to charge money for a ticket, the government would levy some tax on us. That would make it unviable for us. Secondly, the local cinema owners would see us as a threat. They would prevent us from hosting these shows in the vans. There was a fear of being attacked by them, too.

Then I thought of another audacious idea, that of setting up a high-power transmitter on a Himalayan peak in Nepal and beam shows to India, and cover the rest of the country from the waters, as India is surrounded by sea on three sides. I imagined the coastal area and the hinterland could be covered by putting up large barges in the sea mounted with high-power transmitters.

I was so tired of the strange and restrictive regulations that I wanted to operate in places where I would be free of them. Transmitters in Nepal and the seas would keep me away from regressive thinking of state and Central governments.

I contacted an industry leader in Nepal and asked for his help. He suggested that we meet with officials working with the king of Nepal to explore the idea. I travelled to Kathmandu and, with a Nepali friend, went to discuss this idea with a senior official in the king's office.

But the official declined to help. He said it would be impossible for them to help if the Government of India did not allow such transmission. I even met the Indian ambassador to Nepal. He actually laughed at the concept. I had no choice but to drop this idea.

I went back to the drawing board to think of new ways of reaching people's homes. Then, a couple of months later, the Gulf War broke out. It was the beginning of 1991. My friends would invite me to hotels to watch CNN's coverage of the war. Some hotels had acquired satellite dishes to receive signals of CNN in their rooms.

This struck a chord with me. I began to wonder why entertainment could not be shown to the Indian public using such satellite dishes. I started asking around about it. I learnt that the government-run Doordarshan (DD) network was already using a satellite to connect all its centres to beam its shows. I asked friends in Mumbai what a satellite was and how it helped beam shows. But nobody seemed to know anything more than the terms. They only knew it was some great new technology being used by CNN and DD.

Then I thought of a friend of mine from Hisar who was in school with me, Gulshan Sachdeva. He was working in DD as a producer. I contacted him with some difficulty and met him in Delhi.

I explained that I was exploring the possibility of starting a television station with entertainment programming. He did not believe that it was possible by a private sector person as it would require a huge infrastructure. DD had thousands of transmitters throughout the country and it had more than 50,000 employees and equipment worth hundreds of crores.

However, I insisted that I had to do it. My stubborn streak was active. I was restless about doing something new. Once I got a sense of how TV broadcast worked I was excited by it. For me it was even more exciting that no one in the private sector had tried it. It is a cliché, but I did not believe anything was impossible.

Once I was convinced of the need to start a project, I would be dogged about it. I had courage of conviction that went beyond my abilities. For instance, I had started speaking English but I was miserable at it. I would speak in broken English. But that did not embarrass me at all. I felt unless I tried it, I would not improve.

I think a sense of daring also guided my instincts. I was getting ready to get into a totally new activity that had no precedent in India. After my success in bringing laminated tubes to India and launching an amusement park, I was confident.

I had been able to face challenges from global companies like HLL and I had been able to build the country's first privately-owned amusement park. There was no reason why I should not attempt to get into the broadcasting sector.

Here again I would credit Dadaji for my attitude. I would be daring enough to jump into situations that would scare others. I did not fear anything. This lack of fear was the legacy of my Dadaji. He used to say that people feared death most. But death comes only once. Why should anyone be scared of something that happened only once.

I was also a thrill-seeker. If a project did not challenge or thrill me, I could not commit to it. I sought a sense of conquest with every new idea that I adopted.

TO ME, BEING the first entrant in a business was more important than being the last. Many businessmen enter a sector when it is mature and has many players. I prefer to be the first. I want to remain in number 1 or strong number 2 position, nothing less than that. I don't like entering a crowded sector. By being the first in a sector, you can guide and shape its evolution.

The television sector ticked all this categories. It seemed like the perfect sector. I told Gulshan Sachdeva that he had to help me crack this challenge. He suggested that we meet the deputy chief engineer of DD to understand how to proceed. We wanted to ask him how a person like me could do it. We were keen to understand what a satellite was and how it worked, how shows were beamed across countries. I had a broad idea now, but wanted to get into technical details.

I suggested to Gulshan that I should be introduced as a non-resident Indian (NRI) to his deputy chief engineer. He might not have entertained us if I was introduced as a domestic businessman. Also, an NRI was more likely to invest large sums of money in such services.

We met him over tea. He asked many probing questions, some of which I could not answer. I told him that I wanted to start a TV channel and was keen on his advice. He said that using a satellite was the only answer. He was sceptical of my intentions and did not take me too seriously. He said that many people talked about setting up TV channels, but nobody had actually done anything.

I came away from that meeting but met him three or four times over the next few days to convince him of my seriousness. Finally, I could persuade him to make a techno-economic feasibility report for me. He brought in another engineer from DD who agreed to work with him for us. It was routine for many officials and engineers of DD to moonlight for some extra money. Even Gulshan used to produce audiovisual presentations or commercial films for some private companies.

The deputy chief engineer asked Gulshan for Rs 25,000 to prepare the report. To my mind it was a very reasonable fee, much less than I expected. I sent Rs 10,000 through Gulshan. For fifteen days I did not hear anything from them. I chased Gulshan and he chased the engineers. Finally, after about four months, I got the report. It was pretty long. I think it is still in our archives.

The report recommended that we use the AsiaSat satellite. It had an Indian footprint and covered India very well.

I then asked how I could get access to the satellite company. The DD engineers procured the contact numbers for us.

I had no idea about East Asia and the business environment in Hong Kong, where the AsiaSat offices were. This was a dark part of the world for me. But I was not ready to give up and chose the option of making a cold call to AsiaSat.

I drafted a letter of interest and faxed it to AsiaSat in Hong Kong. I wrote that we wanted to hire a transponder of the satellite for beaming shows for India. Not surprisingly, there was no response.

I sent several reminders by fax. I called their phone number but could not get my point across as the girl at the other end hardly spoke English. After about one-and-a-half months of trying to contact them I decided direct action was required. I would have to go to their headquarters. I reached their office in Hong Kong and asked about the CEO of the company. I was told that he was on an official visit to Canada. Luckily, I could persuade the front office to give me the CEO's number in Canada. That night I called his hotel and managed to get through to him. He was sleeping and my call woke him up. He was very irritated. I told him that I was keen on hiring a transponder and had been trying to get in touch for last three months without a response. He replied that he could not offer any transponder as all of them had either been booked by another Hong Kong-based company or it had the first right of refusal on the satellite capacity. I requested him to send me the details.

Next day his office sent me a terse two-line fax with the address and contact details. The company was Satellite Television for Asian Region (STAR) Ltd.

Now I tried to get through to this company for hiring a transponder. When I reached their office, they asked whether I wanted to sell them programmes. I was confused initially but then realized that the company was in the process of launching a couple of TV channels in the Asian region with Western programming. It planned to telecast BBC and MTV in Asia initially. They used STAR as their brand name, an acronym for Satellite TV for Asian Region.

Star TV was owned by local business magnate Li Ka-shing, one of the richest men in Asia. Li Ka-shing owned about 30 per cent of the satellite company. Another 30 per cent was with the government of China and the rest was perhaps reserved for the public. But Li Ka-shing effectively controlled the company. The Chinese government had asked him to lease a transponder to the Burmese government to run their state TV station. He was also asked to provide satellite services to any other government that sought it, presumably with China's approval. But among private companies, he alone had the exclusive rights through his son Richard Li, who owned and managed the Star TV company.

I realized that I would need help to approach this group. A cold call would not help with the Li family or their executives. They were too big and powerful to entertain the ideas of an unknown rice trader and owner of a packaging company from a poor country like India.

14

THE $5 MILLION GAMBLE

Despite an incredible offer,
I am still the last choice!

TO TAKE MY broadcasting project forward, I roped in Ashok Kurien. He was the head of the advertising agency Ambience, which had worked for the launch of Essel World. Advertising folks have a knack of making friends out of clients, and Ashok and I had become friends. He had some experience of buying programmes for Doordarshan. He also told me how ads were being included in movies being distributed on VHS videotapes. The videotapes were sold to consumers so that they could watch the films in their homes on a video player. By then a few enterprising individuals had started showing newly released films in residential areas. They would use a video player and connect it to homes of people with cables. They would charge a monthly fee to show one movie a day and a couple of new movies per month. This was the beginning of the cable TV revolution in India. About a million homes throughout the country were already connected by these operators. Most video parlour owners turned into cable operators.

Ashok reached out to his Hong Kong-based friend, Ranjan Isaac. Ranjan was a partner in an advertising agency in Hong

Kong and knew the CEO of Star group. Through Ranjan we reached out to the Chan brothers—Robert and George—who were senior executives at Star. Robert was the CEO while George handled other operations. Robert deputed a middle-level executive, an Englishman named David Manion, to talk to us and understand what we wanted.

Finally, we could explain to someone that we wanted to launch a TV channel with Indian programming, and hence we were keen to rent a transponder on AsiaSat satellite. David was not sure about our abilities and suggested that we make some programmes to show him. I told him I had some ideas about programming but that I was not an expert. He then asked me to make a showreel of likely/proposed programmes.

We returned to Mumbai and Ashok's team put together a showreel with programmes like *Chitrahaar* (the popular film-songs-based show on Doordarshan), some chat shows, a news capsule and clippings of famous movies.

We returned to Hong Kong and showed the showreel to David Manion. Somehow, he liked the pilot. He asked us if we were open to a joint venture with Star. We could launch a TV channel instead of just renting a transponder. I was happy on one hand, thinking that we would get the necessary expertise to run the business. On the other hand, I was apprehensive that we might lose control of the channel to the Chinese.

After some days, I realized that we had no choice but to accept the idea of a joint venture. Confirming this to David Manion, we told him that we had to own at least 50 per cent share in the JV and that we would be in charge of managing the channel.

Despite's David's positivity, his Indian assistant, Atul Kapoor, was very negative. Kapoor said that the intellectual property rights of the movies in India were very fragmented and getting broadcasting rights from producers would be almost impossible.

Even if we took the rights from the producers, there would be many claimants of such rights, and hence it would lead to long-drawn-out legal battles. During most of the meetings he would criticize India in every way possible. He was especially critical of our entertainment industry. His negative comments about India made me angry, but I controlled my temper. However, on the copyright issue, he was proven right, as we later discovered.

After beginning our talks with David, we spent the entire year of 1990 checking on the legal position and prepared a business plan. The business plan showed a breakeven after three years of launching the service. Profits were likely from the fourth year onwards. The plan included a transponder cost of US $1.2 million per year and a 50:50 JV. We were aware that intellectual property rights were a problem if we were to use movies for our shows. And, of course, there was the critical fact that Indian laws did not allow private companies to be a broadcaster.

I asked lawyers in India how CNN could beam into India. They said that it was illegal for a foreign channel to be shown in India using a foreign satellite. Under the law, broadcasting was an activity reserved for the Government of India or the public sector. Even a state government could not enter this activity. But since the government had not blocked signals of CNN or taken any action against the owners of satellite dishes on ground, the beaming continued. Knowing the possible problems that our venture could face, David Manion and Star executives put the onus on us to sort out legal and regulatory issues. This was part of the conditions of the JV.

I believed that the government would sooner or later have to change the rules. But I did not want to wait for the government to open up the broadcasting sector for private players. Once it opened up formally, several people would jump in. I preferred to start when no one else was thinking about it. It would be tough for the government to stop us, as they were allowing CNN. And stopping CNN, or any other TV signal, would make the government appear anti-progress, I thought.

Still, it was a risk. If the government decided to stop the telecast, there was nothing we could do.

We decided to go ahead with our plan.

After David was convinced, the proposal was taken to his boss Steve Moss. It was Steve who was supposed to take it to Richard Li. After a while Steve gave us an appointment. Steve was of Canadian origin. He was pleasant but seemed to be suspicious. He had many questions for David and me. Neither he nor any other executive in Star had much of an idea about the Indian market. India was not on their radar and not part of the pan-Asian broadcast plan. For Richard Li and his team, pan Asia meant the Greater China region, including China, Hong Kong, Singapore, Taiwan and Korea. Then there was Atul Kapoor and the company's headhunter, Ranjan Marwah, who gave negative inputs on the Indian market to Richard. Marwah didn't know India but depended on a couple of his buddies in the country for generic information. After our meeting, Steve Moss remained sceptical, but promised to promote the idea and concept within Star.

IT WAS 14 DECEMBER 1991 when Ashok Kurien and I reached Star TV's office in Hong Kong. We were greeted by David Manion, who took us to meet Steve Moss. Steve gave us no commitment of any kind and told us in an ominous tone that it was time to meet Mr Li. This meeting was to take place after two days. Those two days that we spent in a Hong Kong hotel were full of anxiety since we faced total uncertainty about our future.

When we reached the office for the meeting, we were escorted to the main conference room of Star. There were ten to twelve senior and junior executives in the room. Richard Li was not there. So we waited awhile. It was like waiting for the king to come in and give his blessings. Richard walked in suddenly and sat opposite me. 'OK, Indian channel...Hindi channel. Where is the money in India? There is no money in India,' Richard was very dismissive. 'I am not interested in a joint venture.'

Most of us in the room, including his executives, were shocked and stunned. This was too abrupt. It appeared that Richard had already made up his mind about the futility of the project. It could also have been a tactical plan on their part to gauge our reaction. We had hoped that the meeting would result in a formal acceptance of the JV proposal. We thought Richard was meeting us as a prospective partner and to quell any doubts he might have had since his executives had done the groundwork. We had been working on the project for about eight months. Now it was all about to go down the drain. When Richard spoke, he did not even address me. He spoke directly to his team of executives. I was very angry for being summoned for a futile meeting. It was now or nothing for me.

So I addressed Richard directly. 'Mr Li, if you are not interested in the joint venture, can you consider leasing the transponder to us?'

Richard looked around at his team. 'What's the cost of the transponder you have taken in this business plan?' he asked. One of his executives said $1.2 million per year.

'How can that be? Where do you get a transponder for $1.2 million? There is no transponder available for less than $5 million per year,' Richard said. It was a haughty statement to put me off. After that he didn't say another word. He just walked out. The total meeting with Li lasted just about ten minutes.

I think he was testing us and trying to send us packing. He was telling us that this was bigger play than we could handle.

I was seething with anger now. I may not have been a global player, but I was still a successful entrepreneur. I might not have known about the TV sector, but I had worked on the project for a year, with great passion. I expected some respect and time from Richard.

From the time I had thought of this idea till the time of the meeting, it had almost been two years' effort (twenty months to be precise). I was losing my patience. I could have walked away from the room and tried again later. But I stayed. All his executives were also sitting dumbfounded in the room. They did not know what to say and where to look. My brain was spinning with thoughts as well as anger. I was also thinking about the fallback steps. I was not thinking clearly. I forced myself to cool down. I closed my eyes and went into Vipassana meditation for a few seconds.

Then I opened my eyes and calmly said. 'This is fine. I will pay $5 million!' It was a spur-of-the-moment decision. I did not realize the implication of what I had said. Raising that kind of money and making the business viable with that fixed cost would not be easy. 'But my condition is that we sign the agreement of leasing the transponder today and right now,' I said forcefully.

The executives sitting in the room could not believe their ears. They asked me to repeat what I said. Then some of them said in excitement, 'So we have a deal, you wait, we will just come.'

I turned to the remaining people and asked them why they had remained silent and not supported the idea. They had worked with me for many months. I had done whatever they had asked. They had visited India, met advertisers. Both sides had helped each other prepare the business plan. But they did not have the courage to speak in the face of Richard's outburst, or maybe they did not want to contradict him.

Ashok was looking at me with total surprise and shock. He was completely unprepared for what I had said and was kicking me under the table. He whispered, 'Subhash, this is suicidal, we will lose a lot at this price.' But I was in no mood to listen.

The executives were flustered and could not decide on this alone. They asked me to wait while two or three of them scurried away. I assumed they had gone to meet Richard. Ashok, I and other Star executives waited, holding our breath. After about one-and-a-half hours, they returned and said, 'We understand your point of view, your frustration. But we are sorry, we can't sign right now. We will get back to you.' My bold gambit had only got me a vague answer. No commitment and no answer. I was very disappointed. I needed time to think about my next move. But I was sure about one thing. I would not give up the idea. I had to think of a way forward.

WE RETURNED TO Mumbai. Ashok was angry with me and sad at the same time, because of the outcome. But he was relieved, too. I could still not understand why they had denied us the deal. They could have made $50 million on a $12-million transponder cost. I was not sure how we would have made money if we gave them $5 million a year. The cost of programming could go up to $10 million a year. Thus there was no logic in my increasing the cost of transponder to $5 million from $1.2 million. It was a gamble for me. *'Jo hoga, dekha jayega,'* we will see how this plays out, I had thought.

The drama was not over yet. My offer to Richard had sparked fresh thinking in Star. Within a few weeks of this meeting, Star executives descended on India. These included David Manion and Steve Moss.

They met the owners of all media houses, including *The Indian Express, The Times of India, India Today* and *Hindustan*

Times. They even met movie producers. They were trying to understand India. They had realized that India was not to be sniffed at. But they did not meet me.

And I found out the reason later. When the executives had left the conference room to discuss my offer, Richard had asked his headhunter, Ranjan Marwah, to check my credentials and capabilities. This had happened while I was waiting in the meeting room with Ashok and other executives from Star.

Marwah had called people in India to check on me. The feedback he gave Richard was negative. He said that most people had not heard of me, that I had no experience in media. Therefore, I would not be the right partner for Star. This feedback was the reason for their refusal to sign the lease of transponder the same day.

STAR BECAME INTERESTED in India but not in me. They realized the potential of India, but were keen on a partner with a media background. I did not have much choice but to stay away from Star. But as I was obsessed with the TV project, I began to look for alternatives. As I had contacts in Russia from my exporting days, I started enquiring about their satellites. I travelled to Moscow in January 1992 to explore options to rent a Russian satellite transponder.

I spoke to some officials and businessmen. Those days Russia did not have a policy to rent transponders. But some of my friends promised to look for ways to help me rent one. I was told that if I gave them a month, they would find some solution for me. They would have worked on obtaining some approvals. My friend Gulshan and the DD chief engineer told me that the Russian transponder was not as good or powerful as AsiaSat. But they assured me that it would work for our purpose. Gulshan said that AsiaSat should be my first priority, but if I wasn't getting it, I should get the Russian one.

Meanwhile, I learnt that Star was not getting anywhere in their search for a partner in India. The media houses they met were neither convinced about the practicality of launching a private satellite TV channel nor would they even consider paying $5 million for a transponder. Star did not want to go below the offer I had made. And none of the media houses could gather the courage to enter the sector. Even *The Indian Express* could not muster strength despite the backing of Nusli Wadia of Bombay Dyeing. More than affording the $5 million price tag, none of them had any concept or vision of satellite television. I returned to Mumbai and hoped that I would be able to get a Russian transponder soon.

Then one day I got a call from Steve Moss; the great Asian giant had finally realized that I was worth their time. I was very angry and I told him so. I told him that I knew about Star's activities in India for previous thirty to forty days and that I was deeply offended. I told him on the phone that I was not keen to meet him or Star again. Still, Steve almost pleaded for a meeting. I was in Delhi that day and he had tracked me down. He was visiting Delhi, too.

Now I decided to play tough with Star, as I had the Russians as a fallback option. I told Steve that I was flying to Mumbai, and that the only way for him to meet me was to take the flight with me.

I told him that Star did not deserve to be treated with any respect and that I would not make any effort to find time for him. So Steve scrambled to be on the same flight as me. We finally talked on the flight. Steve said he had convinced Richard that I was the right partner for Star.

'You have been meeting all the other Indian media houses,' I told Steve.

'But they have no idea about TV,' he said.

'I know that they are not willing to pay you as much as I am.

They don't even have the conviction about the potential of the sector. That's why you have returned to me,' I countered.

Steve was a bit sheepish. But I told him that I would consider reviving our discussion.

Then Steve said that Richard would be happy to come to Mumbai and meet me. I agreed.

Despite Steve's overtures Star was still not convinced about me. Star tried another gimmick. Richard did not come to Mumbai to meet only me, as Steve had said. He met all the other media heads too. He met Nusli Wadia and others on 19 and 20 May 1992. He met me soon after meeting them. Perhaps I was his last choice. He wanted to be absolutely sure that the other media groups were not keen on TV.

When he came to meet me, I gave him a sense of what our business was and how we were in a position to launch a TV channel. I took him to Essel World by helicopter and showed him around the park. Then I flew Richard to the Essel Packaging factory. He noted that our factory was supplying to Colgate, Unilever, Procter & Gamble and many other brands. Apart from tubes we were also supplying other packaging material. He appreciated then that we were not such a small company and that Ranjan Marwah's feedback about us was flawed. We were not to be scoffed at.

By mid-1992, our group turnover was close to Rs 1,000 crores. We had assets of over Rs 500 crores. Some of the biggest MNCs were clients of Essel Packaging Ltd. Richard knew these companies well; they were also some of the biggest advertisers of Asia. This was important from the point of view of earning ad revenues for the TV channel. He thought that such clients would help me pay the $5 million annual fee I had promised.

The visit to Essel Packaging clinched it for Richard. It was then that we signed a single-page letter of intent on 21 May 1992 in Mumbai's Oberoi Hotel.

I roped in a few NRI businessmen as partners for the TV project. These were Ranjan Isaac, who owned an ad agency in Hong Kong; J.D.R. Malhotra, a friend of Jawahar's who supplied foodgrains to hotels in the UK; and Mohan Tolani, a business associate who helped us import pulses from China. These three became the owners of Asia Today Ltd, which launched Zee TV, India's first private satellite channel and among the first three or four in the whole of Asia.

I had brought in these NRIs since I needed money for this project. I convinced Isaac, Malhotra and Tolani about the opportunity and told them that this business would give them good returns. I explained the revenue model to them. I would supply the programmes and charge them 10 to 15 per cent over the cost of production, plus 15 per cent on revenue of ad sales that we generated. They would own the company and the profit would be theirs. Convinced, they put in about $3.5 million. They would own the channel while I would effectively run and manage it for them besides being a content provider as well as revenue officer.

Asia Today Ltd made a simple deal with Star. Richard would lease us the transponder and include the uplinking facility for $5 million per year. He did not go in for a joint venture, but kept an option to buy 25 per cent of the company that owned the Indian channel. This option was valid for a year from the date of signing and starting the channel.

AT THIS POINT a senior executive of Star TV entered the scene. Michael Johnson, a former technical expert with Hughes, had introduced Li Ka-shing to the importance and concept of satellite TV. Johnson had apparently convinced Li that a failed Hughes satellite could be converted into a lucrative broadcasting business. Johnson, it appears, had used Li Ka-shing's resources and his connections with the Chinese government to get the satellite repositioned in space in the correct orbital spot. It had been made operational as AsiaSat.

Li Ka-shing had given the TV project to his younger son Richard. Johnson put together a deal for Richard to bring BBC, MTV, and Prime Sports to his company for broadcast in Asia. An African-American, Johnson was never to be seen in public during 1991-92. I heard later that he was allegedly a fugitive of American law. Some people in Star believed (rightly or wrongly) that Li was providing him protection in Hong Kong.

Johnson started helping us on technical matters. We registered the company in British Virgin Islands with its operations in Hong Kong.

15

IT'S SHOWTIME, FOLKS!

A quiet launch but a new approach to producing shows

JUST A FEW months before Rajiv Gandhi was assassinated while campaigning for the Lok Sabha elections in 1991, a friend from Mumbai, Manubhai Desai, took me to meet him. Manubhai wanted me to contribute some money to the Congress party. Belonging to an old business family from Mumbai and Gujarat, Manubhai was close to Rajiv. When in Mumbai, Rajiv would stay with Manubhai.

Manubhai was surprised to see that Rajiv and I knew each other. Rajiv was warm towards me and scolded me for not keeping in touch. He asked me to give the election contribution to Sitaram Kesri, which I did later.

This meeting renewed my relationship and contact with Rajiv. I met him more frequently after this. We talked about the old days in our meetings.

One such meeting took place while I was struggling to put together money for the satellite TV venture. I think the stress showed on my face. Rajiv asked me why I looked tense. I told him that I wanted to start a TV channel. I explained to him how I would create 50,000 ambassadors for India, and for him (as we

My grandfather Jagan Nath Goenka, the family patriarch who was my idol and inspiration and who had a great influence in my upbringing.

With my father Nand Kishore Goenka and brothers Ashok, Laxmi and Jawahar, who have stuck by me through thick and thin.

At Mandi Adampur in Haryana, the town that my forefathers founded
and where I spent my childhood.

At the Chandulal Anglo Vedic (CAV) High School, managed by the
Arya Samaj community, which I joined in the fifth standard in Hisar.

I got married to Sushila on 4 December 1973 in Amritsar. I was twenty-three, and there were more friends than family at my wedding.

With Sushila in Mumbai, where we now live.

Acchey Din: By the time I was twenty-four,
I had tasted some success in Delhi. It was a
stable and enriching period in my life.

Singing songs at a New Year's eve party.

Posing for a picture with my father, Nand Kishore, and brothers Laxmi, Jawahar and Ashok to accompany a profile on me by the *Wall Street Journal* in 1999. At that time, with the Zee stock skyrocketing, I was being spoken of in the media as the 'richest Indian'.

With my sons Punit (*left*) and Amit (*right*).
While Punit is in charge of Zee Entertainment,
Amit takes care of our infrastructure business.

With my daughter Pooja, to whom I am close.

Sushila and I with our grandchildren.

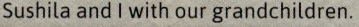

With Priyanka Gandhi at the inauguration in 2009 of a unique pagoda we built for the Global Vipassana Foundation at Essel World in Mumbai.

Receiving the Ernst & Young Entrepreneur of the Year award in 1999 from Dr Manmohan Singh.

It was my ritual to meet Bal Thackeray once every two to three months. If there was a delay from my side, the Shiv Sena chief would himself call me for a meeting.

Congratulating Pranab Mukherjee a week before he became president of India, at a Zee business awards event.

At the inauguration of Water Kingdom at Essel World with, among others, Bal Thackeray and *(to his left)* the then chief minister of Maharashtra, Manohar Joshi.

Honouring former US president Bill Clinton for his initiative to help people affected by an earthquake in Gujarat in 2001. *(Inset):* Myself and my brothers Jawahar *(extreme left)* and Laxmi *(on his left)* with Hillary Clinton.

Receiving an honorary doctorate of business administration
from the University of East London in 2013.

With Amitabh Bachchan and Subrata Roy of Sahara at the
Zee annual bash in Mumbai in 2011.

Receiving the Emmy Directorate Award in New York in 2011.

A ghazal evening at my house with *(from right to left)* Jagjit Singh, Ila Arun, Anup Jalota and Talat Aziz, among others.

With Paresh Rawal and Subhash Ghai at a function in Mumbai to mark 150 years of India's First War of Independence in 1857.

At a special show of Zee's music show *Sa Re Ga Ma Pa* to usher in the new millennium in 2000.

With Pakistani players Inzamam ul Haq and Moin Khan at the first event of the Indian Cricket League (ICL) in Chandigarh in 2006.

A 2008 cover story in *Businessworld* on my efforts to sustain the ICL despite being ostracized by the cricket establishment.

With Tony Greig who was advising us on the ICL.

With my Vipassana guru Satya Narain Goenka, who had a major impact on my life.

Taking Buddha's relics received from Sri Lanka to the top of the pagoda *(right)* built by us for the Global Vipassana Foundation.

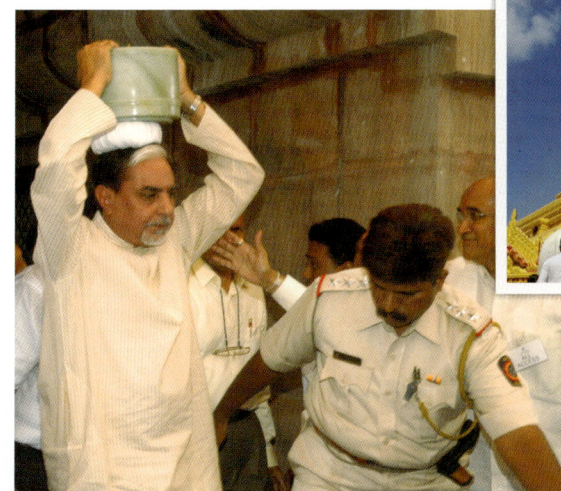

Stabbed in the back: Arriving at the crime branch office of the Delhi police for questioning in a case filed against Zee by steel magnate and Congress MP Naveen Jindal in 2012.

Back to the roots: At the request of the people of Hisar who wanted that I help them change their MLA, I supported the BJP candidate from the area in the 2014 Haryana assembly elections. Seen here with Narendra Modi during a campaign rally.

were hoping he would win the elections). Everyone who watched the channel abroad would learn more about India.

'Why are you worried then,' he asked. I said that I was falling short of money to start the operation. I told him that the people who owned the channel that I was to manage were short by $400,000. He heard me but did not react. But a few days later I got a call from an Indian who said he was speaking from London. He offered to invest the same amount of money. I got the money soon after. Even today I don't know who this person was. This money was critical and it had helped me reach the figure of $5 million required to start the operation.

I had already started raising money for the operations by roping in some private equity and venture capital funds from Hong Kong. One of the funds was owned by Kerry Packer and Sir James Goldsmith. Finally, the three original investors owned 51 per cent of the company while some private equity investors held the rest. In total, we raised about $8 million for 40 per cent equity. After paying for the transponder we used the rest of the money for working capital and running the company.

I did not hold a single share in Asia Today Ltd. Under government rules, Indian nationals were not allowed to buy or hold a share in a company registered abroad. In a way, I was working as a professional for the company owned by my friends. I was earning from the programming cost and advertising revenue share. This was an awkward way to run or launch a business, but there was no choice. The investors had faith in me and in the revenue model. They also believed in the growth of the Indian market.

After raising the money, I started work on branding the channel. I asked my creative team, managed by Ashok Kurien, to come up with a name for the channel. The name of the production company that made the programmes would be the same as that of the channel. Even though they were separate entities, they were bound together. The only person common to both was me.

Elsie Nanji, part of the creative team, offered many suggestions. In the end we went with Zee as the brand name for the TV channel. We changed the name of an existing group company in India to Zee Telefilms Ltd.

Ashok Kurien's Ambience Advertsing had helped me plan for programming and also created the name. He also helped generate advertising revenues. I gave him about 5 per cent stake in the company. The rest was with us.

The launch of the channel was both a happy and sad day for me. It was 2 October 1992. Ranjan Isaac, Tolani and I were in Star TV's uplinking centre in Hong Kong. I took the tape and put a *tilak* and *swastika* on it. We folded our hands and prayed for a few seconds. The operator put the tape in the machine. The moment the tape started playing and the machine uplinked it to the satellite, the first Indian private channel had begun to beam. There was no celebration, nor any assembly of friends and family. There were no applause. It was a quiet birth of a child (destined to become a gorilla one day).

I was delighted, as my obsession of launching a TV channel had been realized. I had promised myself that I would bring entertainment to the homes of people. I had to battle obsolete laws and an unhelpful government.

I was sad also that I had to launch an Indian channel from Hong Kong. Technically, I had launched an illegal channel. It was an illegal business, as Indian laws did not allow what I was doing. But this was the future. Someone would have done it sooner or later. I was glad it was me. It was a very lonely launch. If I had done the launch in India, I would have arranged a great inaugural event.

As soon as the tape played, I called my home in Mumbai and asked what the channel reception was like. The first show was a collection of songs from Hindi movies. My driver picked up the phone. The entire house was watching the launch together. He was excited and told me in Marwari, '*Babu, bahut achho lag rahyo hai*' (The show was looking great).

And thus Zee was born.

I WAS EFFECTIVELY the booking agent for advertisers for Asia Today. The money would be paid directly to Asia Today. Only companies with export incomes could advertise on Zee. Every bill had to be routed through the Reserve Bank of India. It was quite a cumbersome procedure.

We used to send the tapes physically by flight through the export channel. The tapes would be taken by flight at night. They reached Hong Kong the next morning. Then the show would be beamed the same evening or the next day, after going through checks in Star TV's compliance department. The shows had to conform to Hong Kong laws and programme code.

Creating programmes was another challenge. There were only six producers in Mumbai who used to make television programmes. These included Prem Krishen, Dheeraj Kumar, Ramanand Sagar and B.R. Chopra.

All of these producers asked for Rs 3 lakhs for half an hour of programming. This was the price they were charging DD. But this price was too much for me. I argued with them to reduce it for me. I said that they had to incur extra costs to get programming approved for DD as well as get the orders, plus they had to travel to Delhi frequently.

I calculated that the actual cost of a show would be less than Rs 1 lakh per half-hour episode. But all of them ganged up and refused to reduce the rate for me. They came for the inaugural party. They wished me luck but said 'more than good luck you need us'. I just smiled and thanked them for coming.

The words of my friend and DD producer Gulshan Sachdeva, that 'it is close to impossible to get four to six hours of original programming', continued to echo in my ears. In the launch show, I addressed the viewers and much of my anger towards the system was expressed in it. I said: 'We are starting a TV service with two hours of programming. But slowly we will increase the number of hours and improve the quality. I can tell you and promise that you will decide our future programming. You will decide which programme will run and which will not. For your information, there are five to six producers who say that that they are the only ones who can produce programmes for your viewing, and hence we have to buy programmes from them. But I think with your help, we will create a lot of new producers and they will produce better programmes for you. Write to us about programmes you like or don't like.'

I started spending all my time watching TV—and pilots of programmes. I would watch every commercial show of DD each day except when I was travelling. These shows were good, but I wanted to offer even better shows on Zee.

I wanted better quality but at lower costs. The only option was to create and promote new producers. We had to find a way to spot them. Ashok brought in a creative professional from advertising, Karuna Samtani. She convinced some professionals who had created ad films but not TV shows, to make programmes for us. Then we got Ronnie Screwvala, who used to make ad films. We commissioned Ronnie and his team to do four shows.

The word about the channel was spreading through trade magazines. We invited new and young talent to produce shows. We met almost everyone who walked into our offices. In the first few meetings, my colleagues and I would hear storylines from them. If and when we liked the story we asked them how much it would cost to produce it. Most said Rs 1 lakh or so. We agreed by issuing a letter of commissioning on the spot. This generally elicited a 'I-don't-believe-this' reaction. After receiving the letter, as they were leaving, they would hesitatingly stop at the door and return to tell us that they did not have the money to produce the pilot episode for the show.

Initially, I gave them an advance for two episodes. We would give them a cheque on the spot. We had to support the newcomers to break the old producers' cartel. It was a risk worth taking and it was necessary. Most of us were not experienced in the field of producing TV shows. Even the producers were learning everyday.

We used to ask the prospective producers to fill up a cost sheet. This sheet would have costs of artists, equipment, etc. We would also ask how many shifts of work each episode would take. We began high-band analogue production, though the prevalent norm was low band. The high band became digital and later high definition.

I visited the sets and locations of some of the shoots. Sometimes I would learn from the crew that the equipment being used was of a lower quality than what we had commissioned. Some producers used a mid-level 'thatha' camera while they were supposed to use a high-band camera.

I embedded myself in the sets to learn the craft of shooting episodes. I would chat with the crew and technical team to understand how things worked. This was thanks to the training my grandfather had given me, that one must know the minute details of any business or activity.

While writing this book, I have to admit that today's system of production has changed a lot. Last year, I went to four to five sets where our shows were being produced. It saddened me. When we started in 1992-93, the entire team of producer, director, actors, technicians could be seen on the shoot. There was a positive and exciting atmosphere. Now it has become a mechanical process and there is no life on the set. The role of producer has been reduced to liasing with the TV networks or to organize funds for operations. The producer is rarely on the set and is not usually involved creatively.

The writer offers the script of the day but is not sure when it will be recorded. There are multiple directors for the same show, based on availability. The artistes drop in, mouth their dialogues and move out. There is no sense of team spirit or camaraderie. The process seems professional but appears soulless.

May be this is the right way. But I believe that the producer, director and writer are the key creative people and must work as a team. They must be on the set and work with the actors and the technical team. They must be involved to ensure that the emotions are being transferred to the viewers. We used to incentivize producers for good ratings and viewer response. Now it is a clinical effort.

In the initial years, the actor stayed with the show till it was over. Now actors are in multiple shows on multiple channels. When Zee started, we preferred new faces and gave them opportunities. Now agents of artistes cut deals with programming teams of channels to ensure work. Newcomers find it tougher to enter and have to pay a high level of commissions to get a slot in a show.

Till 1998-2000, artistes and producers were associated with channels. They were a common team aligned together. New launches were celebrated together and the entire team worked and lived like a large family. Now those bonds are broken.

Let me narrate a story of how we worked. At the beginning of 1993, a young video editor came up with an idea for a singing competition show. It was based on the party game Antakshri, where each subsequent contestant has to sing a song that begins with the last letter of the previous song. He gave this concept to Zee's head of programming. This young video editor was rebuffed

and told that he was not qualified to produce or direct such a show. The editor came to me and asked for a chance to prove the idea. I have always believed in supporting new talent and ideas. So I okayed the project. This show was launched soon as *Antakshiri* and turned out to be among the most successful shows in TV history (subsequently, the same person produced and directed a singing contest *Sa Re Ga Ma Pa*). It spawned many copies but has not been bested yet. I wonder if young talented people these days get a chance to prove themselves like they did earlier.

I ALSO WENT to movie producers like Prakash Mehra and Subhash Ghai and told them I was launching a TV channel. I said we needed help from the movie industry, which in turn needed support from TV. A partnership would help both. I would show their old movies and promote the upcoming one on our TV channel. This would also increase the attendance at movie theatres. I said I would share advertising revenue from the movies I showed on the channel. This was a powerful argument for both industries. But the leading film-makers of India did not agree. All they wanted was to sell their library of movies. They did not have faith in the future of TV to go for a revenue sharing partnership.

I decided to buy the movie rights. I said I would buy a movie's rights for Rs 10,000 to Rs 50,000 for ten years. The movie producers agreed. We bought almost 3,000 movies. One year later, when I visited the movie producers' association, they begged me to reduce the duration from ten years to seven years. I agreed, but regretted it after a few years as the producer fraternity had no gratitude of any kind. Their motive had been short-sighted—that of making immediate profits.

Meanwhile, a new character turned up during my attempts to source content from Bollywood. One Dhirubhai Shah said that he had bought the video rights of movies of all kinds. And since broadcasting was done by playing video, he claimed he had the rights for broadcasting, too. I would have to pay him rather than the producers. Shah said that the producers did not have any rights anymore and even showed me the agreements he had signed with the producers.

The agreement was for video rights and all electronics rights. It did not clarify anything more. Technically, he had a valid argument. Also, the way these agreements were written, the producers could not have won the case. I was in a dilemma about what to do.

On one hand, I could have done a deal with one person at throwaway prices for almost 80-90 per cent of the Hindi movies. On the other, my conscience did not allow me to do this. The producers had signed an unfair and one-sided agreement without realizing the future impact of satellite TV, internet or other technological developments.

Hence I decided to support producers. Surprisingly, the producers were being meek and did not want to challenge Shah. But I persisted and said that the rights were for video rentals on VHS tapes, or for home viewing. In my view Shah did not have broadcast rights.

After I took the broadcast rights from the producers, Shah filed many court cases against us and the producers. His argument was so good that he almost won in the Mumbai High Court.

He did a smart demonstration to convince the judges in the court. He took a VHS tape and a VCR and played the movie on it. Then he asked how we would air the movies. Our lawyers said we would also play the movie through a video player. Shah looked triumphant and said that this proved that the broadcast rights were also owned by him. He convinced many people that playing a movie on the video player was the same as broadcasting it.

I had to get a battery of lawyers to defend our rights. This was the first case of its kind in India. The broadcast rules were not clear and the fate of our case would influence the future of broadcasting in India. We tried to show examples of laws in other countries where the satellite rights were different from video rights. We made a strong pitch to prove that playing of a movie on a video system was significantly different from playing a movie for broadcasting. Video rights meant the distribution of movies on physical video tapes. Broadcast was distribution from a totally different platform. There was no physical distribution of tapes. A movie signal was being beamed through satellite to millions of people simultaneously.

While this seems obvious now, it was a totally new concept in the early nineties. We were arguing that each method of distributing a movie and bringing it to viewers should be seen independently. The court saw merit in this and we won the case. This benefited the movie industry tremendously. Movie producers could now earn revenue from a totally new stream. This would end up changing the business model of earning profits from movies.

Thanks to this case that I fought, the producers were also released from the tyranny of Dhirubhai Shah. We went on a buying spree. The movie producers had never anticipated that they could get more money for old movies. It was easy money for them. Even though I would pay them six to eight months after airing the movies, they were happy. Soon, many producers were approaching me to buy their movies. Sadly, most movie producers did not give me credit for creating this revenue stream nor did they communicate their gratitude to me for winning their case.

WE BEGAN ATTRACTING an increasing number of viewers. But even after two months of the launch of the channel, we were not getting much advertising. My brother Laxmi called me from Delhi one day saying that he saw the previous day's transmission for three hours and found only two ads in it. He was worried that we were not getting enough ads. Actually, we were not earning any revenue. 'Even those were free,' I told Laxmi. Ashok Kurien was working hard on getting ads but the advertising industry and the clients were unsure about spending money on our channel. They were not sure whether we were being watched at all. There was no system of measurement or viewers' feedback. For marketers, the advertising options were either DD or video movies. Without any measurement, it was tough to justify our rates. We knew our viewership was more than that of video rentals. And also that DD was the omnipresent network. We took the most practical step in the circumstances. We priced our ads between these two categories with a rate of Rs 10,000 per ten seconds.

It was a tough sell to advertisers in the absence of any system or proof of viewership. My relationship as a supplier of tubes to advertisers like Hindustan Lever and Colgate did not help. Ironically, Richard Li had assumed that this relationship would help us earn revenues. Each of these consumer companies and their media buyers would ask for viewership numbers that we did not have. DD did not have such problems, since it was available in all TV homes.

Finally, we were rescued by our viewers. Thousands of viewers began writing to us. We started getting letters from viewers complimenting us and sharing feedback on the shows. In an era before email and sms, the committed viewers of Zee were eagerly sending letters and postcards to share their views on our shows.

We were desperate to prove that Zee was being watched by hundreds of thousands of people. When we started getting the letters, we hit upon the idea to use them as viewership data. We took bundles of these letters to advertisers and media buyers. We asked them to study and read them exhaustively to ensure they were genuine. Thankfully, many of the people who were in these companies were also getting personal feedback from their families. Sackfuls of letters proved to be rudimentary but effective viewer feedback data. If we had been smart in marketing, we could have sought recognition from the Guinness book of records for receiving the highest number of letters from viewers. Slowly, the advertising community became convinced of the success of the channel and the revenues started trickling in.

By now some newspapers also started to write about our shows. TV critics reviewed Zee's shows and compared them with shows on DD. The most popular show in the first few months turned out to be Zee Top Ten, the listing of top songs of the week hosted by Archana Puran Singh. The show was produced in a corner of the office. We had no studio for the show. Her enthusiasm on air was welcomed by viewers who were tired of DD. Zee was a welcome break for viewers starved of choice. Archana would get only about Rs 2,000 per episode. Shows like Top Ten helped us establish the channel and go deeper into the mind space of viewers.

We also took on a young Smriti Irani as a *sutradhaar* for the channel—she welcomed viewers when the evening programming began and said 'good night' at the close. It was her first job; she was paid Rs 5,000 or so per month. Today, of course, she is the Union minister for human resource development.

DURING THIS PERIOD we also helped create about thirty new TV producers, including Tony and Diya Singh, Jeetendra, his wife Shobha and daughter Ekta Kapoor, to name a few.

Zee gave many firsts to the viewers. Though seemingly these issues were a reflection of the reality in society, some of them were taboo. People did not even want to talk about them, so showing them on air was not easy. Issues such as extramarital relations and single motherhood were brought by Zee to Indian living rooms. I was criticized for legitimizing unhealthy social practices. When we started a 24-hour news channel, I was accused of corrupting the Hindi language by purists. They were upset that our news shows were in a mix of Hindi, Urdu and English.

We were learning almost every day. We did not have a buffer of shows and as soon as the shows were ready, they were aired. I remember the tragic terrorist attacks and multiple blasts in Mumbai. The city was under curfew for many days. The production of shows was affected as was all other work in the city. I had to make personal arrangements for the team to come to work. Most people were scared of being out on the roads. I had to personally drive to the homes of some staff members to pick them up from home. Even my driver refused to go out in the atmosphere of fear. We succeeded in feeding the broadcast centre with programmes on a daily basis. Very few shows were repeated for want of fresh content.

WHILE ADVERTISING WAS picking up slowly, our investors were getting restless. Even though Tolani, Malhotra and Isaac were happy with the channel's operations on the face of it, their investment in Zee was not earning them anything. They did not see a single dollar in their bank accounts for the first six months. Even the private equity funds were asking me and the Zee team tough questions.

It was a bit disheartening. I was keen to celebrate the successful though nascent steps of the first private Indian channel. The investors, however, ignored the pioneering work and fretted about their payback. I had an argument with the investors. I told them to be a bit patient as we were doing our best. The revenue generation was taking longer than we had thought. We had been on air for only six months. But none of the investors was convinced about the future prospects of the channel. They liked the idea but sought quick returns on their money.

The three individuals who had invested before the financial institutions did, wanted an exit. They asked me to buy them out but I did not have the funds, more so since I would have had to pay in foreign exchange and not Indian rupees. The only thing I could buy was time.

However, to ensure their commitment, I agreed to buy them out, though with a caveat. I said they could exit but would get a return on their money only after two years. I guaranteed to return their money with appropriate interest. Thankfully, they accepted my terms and I signed an agreement with them. Now I had to face another issue. I could not officially own a company based in Hong Kong. Indians were not allowed to buy or own foreign companies. I took the only way out; I became a non-resident-Indian. To buy and officially own the company I had launched, I had to leave my country. I took up residency in Hong Kong since the company was operating from there.

I became a non-resident Indian and then bought back the three investors' stake in the second half of 1993. I had to live in Hong Kong and stay away from India for a minimum of 186 days in a year to maintain my NRI status. Many years ago, I had to leave my hometown Hisar to succeed, going first to Delhi, then to Bombay. Now I had to leave India to climb the ladder of success.

While my original investors had a pessimistic view of the channel, Richard Li did not. He started getting great feedback on Zee from viewers and industry partners in India. He woke up to the growing market—from about 1.2 million homes at the time of Zee's launch in October 1992, our reach had increased to an estimated 8 million in a year. Richard asked his team to do some research. The research endorsed our success and projected greater growth for the channel in the following years.

The high growth prospects of Zee led to a 180-degree change in Richard's attitude towards us. He contacted me and reminded me of the clause in the contract that allowed him to buy 25 per cent stake in Zee. He had had his cake by charging four times the rental of a transponder from us. Now he wanted to eat it as well by acquiring Zee. He was keen to partly own a company that once he did not want to even consider working with.

Sadly for him, this realization dawned on him too late. When he checked the agreement, he realized that the deadline for exercising his right was over. He had an option to buy a stake in Zee within one year of signing the transponder agreement. By then more than a year had passed.

This was a small setback for Richard but he persisted and tried hard to persuade me to sell a 25 per cent stake to him. There was now a realization that India would be an important market, and that Zee would be the key to making it big in India.

I refused. Now I was successful and did not really need him. And I wanted to run this company as it was mine now. Our team had made it a success.

After buying the stake from the original investors, I was the owner and in control. The institutional investors were allowing me to lead the company in my own way. Within two years of the launch, Zee was successful in the eyes of investors, viewers, the business community and the government. We had spent more than Rs 60 crores for the launch and programming in one year of operation, though after the first six months of the launch, we earned just Rs 2 crores from advertising revenues.

But there was an inflection point after over one year of broadcasting. Suddenly, our advertising revenues shot through

the roof. We went from Rs 2 crores in first six months to Rs 100 crores in the next twelve months. My belief in broadcasting had paid off. After months of struggling and almost begging people for a chance to launch a channel, I had been able to make a success out of Zee. Companies like Star TV, which had longer and deeper experience in broadcasting, were now appreciating our vision. I had been able to overcome reluctant satellite partners, impatient investors and sceptical advertisers. The faith shown in Zee by the viewers had overcome every doubting person and institution.

16

WHO WILL MANAGE THE MANAGERS

Renewing teams to keep up the pace

LAUNCHING A BUSINESS has one dynamic, but running it has another. In running Zee, four senior members supported me. Though I was the CEO for the first two to three years, the working environment was such that all the core members of the launch team considered themselves CEOs. They included: Ashok Kurien, the stakeholder/co-promoter who helped in different functions; Digvijay Singh, in-charge of revenue; Karuna Samtani, who managed programming; and Gopal Maliwal, the finance head.

After the first eighteen months of operation, I requested Digvijay Singh and others to prepare the budget for the next year on both revenue and expense accounts. So far we were in survival mode. Any revenue was good revenue. After settling into a rhythm, I wanted a medium-to-long-term growth plan for Zee. My experience so far had taught me that unless there was a clear path that matched costs and earnings, any business would collapse into lethargy and ruin. Singh and his team worked on it and came up with a 20 per cent increase in advertising revenues, as against a projected 15 per cent increase in overall

advertising spend of India. At the same time, they said that the costs would rise by about 30 per cent over the previous twelve months.

The spurt in advertising that we had seen did not reflect in the assessment by Singh's team; I was surprised by their conservative approach to growth. Such an approach is acceptable when the business has matured. But they were forgetting that the first six months' revenue of Rs 2 crores became Rs 30 crores in the next six months and a further Rs 70 crores were added over the following six months.

It was too early for us to slow down and accept a moderate growth rate. I realized then that we needed a new team that would look at life at Zee with a fresh approach. I did that and was proven right as Zee grew much faster with a new team.

I am of the view that in any business whatsoever, the start-up team of any business should move to another start-up. They should not be expected to manage a running business. In the beginning, everyone in the start-up mode is hungry for success and does everything possible to succeed. After the initial success, complacency sets in. The same happened with Singh's team. They wanted to take it easy though I felt that the same growth rate could be there for another two decades. I moved Singh's team to other projects and ideas while I set about looking for a new one.

I asked B.K. Sanyal, who was on the board of Essel Packaging and part of its many active committees, to hold the CEO's position at Zee till I could find a regular CEO. This worked very well for the business as he was greatly analytical. He could provide a greater insight into the growth of the business. Meenakshi Madhwani, an advertising professional, joined us as revenue head. Sanyal was a straight-talker and held a mirror to other team leaders. I got along well with him, too. We also started one-hour band of Bangla and Tamil programming on the Hindi channel. This experiment was successful and paved the way for the launch of a 24-hour Zee Bangla channel.

We would have launched a Tamil language channel as well, but by then I had asked Vijay Jindal to take charge as CEO. Sadly, he shot down the plan for Tamil channel, wanting to focus only on the flagship. This opportunity was grabbed by the Sun TV group, led by the Maran family of Chennai. They were persuaded to do so by Zee's former head of Tamil programming.

Jindal was brought in because matters had begun to sour with Sanyal. Sanyal felt that our revenue was not comensurate with our ratings. He also pointed out many other weaknesses in different sections. This did not go down well with the senior team members. Sanyal would give senior executives like Digvijay Singh, Meenakshi Madhvani and Karuna Samtani honest and brutal feedback about their work.

This created a lot of resentment against him. These executives were so upset that they all got together and walked into my cabin and said, 'We have all reached a point where either we will leave the company or Sanyal does.' I had to make a tough choice. Instead of losing all other executives I had to request Sanyal to vacate the CEO's position. I told him, 'Sanyal sahib, what you say is right. You are speaking the truth. But this means these executives will not stay. People in the trade and in their circles call them heroes and they are perceived to have put in an excellent performance. They feel that they created Zee TV. But you show them a reality that they can't face.'

Sanyal said, 'Subhashji, I can't change myself.' So I asked him to join the board and supervise the group from a non-executive position. I had to speed up the search for a CEO. Sanyal stayed on the board of the company till his death.

Vijay Jindal was brought in as CEO. Jindal was working for *The Times of India* group. Jawahar had met Jindal for some matter when he was in *The Times of India* and we thought he

was a good choice. We asked him to join Zee as CEO and signed an agreement with him. But I did not realize how critical he was to the promoters of *The Times of India*. When Jindal submitted his resignation there, the Jains, owners of the newspaper, reached out to me through their friend Murli Deora, then president of BRCC (Bombay Region Congress Committee). Deora would later become a Cabinet minister in the Central government. I received a call from Ashok Jain, promoter of *The Times of India*, late at night when I was in Hong Hong. I was in deep sleep when the call came. I was requested not to hire Jindal. I told Jain that we had already signed an agreement. However, if Jindal wanted to opt out of his agreement with Zee, he was free to do so. I told Jain to talk to Jindal about this. If Jindal changed his mind, I would not pursue the matter. However, Jindal was convinced about joining Zee and did not change his mind.

Along with Jindal, several senior executives joined. Nitin Keni was given charge of programming, and Hitesh Vakil was given charge of finance and accounts. Meenakshi continued as the revenue head. With the new team we doubled our earnings that year while the costs remained manageable. This was far better than the projections submitted by the start-up team led by Ashok Kurien and Digvijay Singh. In fact, we have not looked back since then. I have yet to see a business growing at 30 per cent a year for twenty years and showing no signs of slowing down. However, in all this time, there has been constant infusion of fresh talent. The company has had a total of seven CEOs in twenty years.

I had withdrawn from day-to-day operations and decided to work from a different location. We had no other office in Mumbai, hence I moved my office to the Oberoi Hotel. I had been advised that if I wanted Zee to become an institution and last beyond me, I should distance myself from its day-to-day working and hand over the running of the group to management professionals. Vijay Jindal had made this presentation to me and convinced me about it. I realized it very late (after about twelve months of my withdrawing from day-to-day management) that this might have been a ploy by Jindal, so that he could control the group without the supervision of the promoters—that is, myself and my brothers.

Jindal seemed to have started playing one brother against the other. Over a period of time, he created lots of differences between my brothers and me.

I heard later that he had a troubled childhood. His parents were old when he was born. The age difference between him and his elder sibling, who became a businessman, was eighteen to twenty years. Jindal seemed to have developed some kind of contempt against people with money or those who had authority over him. This was my conjecture after several interactions and discussions with him.

Vijay Jindal would feed me negative news about my brothers. And to my brothers he would say I did not like their work. Once he even told Jawahar that I did not want him to look after a specific division, though I had taken no such decision. All of us brothers were in touch always on issues of the day or the week. But when we met each other, we did not always talk business. My brothers had a lot of respect for me, and accepted my decisions. Even if I did not tell them about something directly, they would accept it when people like Jindal quoted me. Naturally, my brothers started getting upset.

Jindal seemed to have created an environment of mistrust. My brothers were made to feel that I did not want them to be involved in the media business or in decision-making in general. The media business attracts everyone's attention. And there were hundreds of requests by family members and others—for coverage of some event or the other, or help in serials. But Jindal would not allow such requests to be met. This hurt all my brothers, who rightly felt that they could not get things done in their own company. While professionals are important, the views and interests of shareholders can't be ignored. A balance was needed. Either way, it was important to be transparent in such decisions.

Despite his attempts to create a divide between me and my brothers, Jindal did help the company. He was a brilliant thinker and an excellent executive. He delivered the top and bottom line for the business. I could give him an assignment and somehow he would deliver on it. We started our overseas operations by buying out TV Asia UK, a general entertainment channel, for about £5 million, and also started operating Zee TV in the USA as a greenfield project. Jindal was of the view that one should concentrate on the flagship brand. That's how we changed our 'flanking' TV channel, EL TV, which had been launched to protect the main Zee channel from competition, into a news channel. Though Jindal was not much in favour of regional channels, I continued my expansion plans in this area. Jindal was also good at fighting legal battles. He had brought in advocate M.B. Zaidi to help him; Zaidi stayed with us for a long time, even after Jindal departed.

I was regularly in touch with our investors. When we were buying out TV Asia UK, the minority shareholders of Zee Telefilms felt that this buyout might be risky, hence they suggested we do the overseas business on our own. Till this time the listed entity Zee Telefilms Ltd had three revenue streams: one was the sale of programming to Asia Today Ltd and the mark-up of profits on the same; the second was advertising sales commission; and the third was receipt of payment from programmes sold to overseas markets. This was at the time of the first tech boom worldwide in internet, media and technology shares. Zee was part of that upswing.

I was finally able to take Zee shareholders' approval to buy

ATL. As a result ATL became a 100 per cent owned subsidiary of Zee. Jindal played a key role in all these activities. During his tenure, Zee's stock rose exponentially, mainly due to the tech boom but also partly due to his and his team's performance.

After a while, Jindal seemed to lose his enthusiasm for work. It was as if he wanted to leave. Why would a CEO who was getting all that he wanted leave the job? Having seen the figures, I know that he had earned more than Rs 60 crores by selling all the shares of Zee that he got as stock options. After making such a huge gain, he perhaps had no incentive to work any more. It was not only him; Jindal had recommended that stocks be given to two dozen individuals and the board had agreed to his recommendation. But soon, these people became complacent and stopped performing. Many of them either left their jobs, or became non-performers. This left a very bitter taste in my mouth. The way people misused stocks options was deplorable. We did not start any other stock option plan for any business for the next ten to twelve years.

A FEW MONTHS after Jindal left, our main Hindi entertainment channel, Zee, started losing its number one position and slipped to third position. However, as a network we remained steady at number one or number two position. One of the reasons for the slippage was the launch of regional language channels, which were not as profitable as Hindi channels.

I had to remove Jindal's successor as CEO, R.K. Singh, as he could not stem the fall in ratings of the flagship channel. He appeared to be working with a coterie that surrounded him. While Star was becoming a strong competitor with the success of *Kaun Banega Crorepati*, my professional managers were not delivering.

17

THE NEWSROOM

*A change in my profile as I get
into current affairs*

IN 1993, I along with Nitin Keni, head of Zee TV programming, visited MIPCOM (a global trade exhibition at Cannes in France), to look for programming ideas. This was my first such visit. We ended up buying some Latin American soaps. We needed fresh content and I was looking for cost-effective options. The characters in these serials had black hair and dark eyes like us Indians. We aired these shows on Zee TV with Hindi dubbing. These did very well and viewers as well as advertisers liked them. We had to resort to tactics like these to keep freshness in our content. The content being generated by producers was not enough to meet the demands of our growing channel(s).

Those days there were hardly any Indians at MIPCOM. A chance meeting with another Indian at MIPCOM led to the eventual launch of our news channel. He was Sanjeev Prakash, owner of Asian News International, a TV news agency in India. ANI gathered news content on India and supplied it to global news networks. While chatting with him, I thought: why not have a weekly news round-up on Zee? Since there was no private network in India, it would make sense for us to have news on Zee. Prakash agreed almost immediately. His agency was gathering electronic news with camera teams across the country. ANI recorded visuals and sold them to DD and many overseas media companies. ANI could expand with our association since we needed content from across the country. It was too early for us to set up news bureaus, and so I saw value in associating with ANI.

We were not sure whether there would be enough content or interest in a daily news show. So we decided to begin with a weekly news bulletin, and called it *Ghoomta Aina*. We agreed to pay ANI Rs 15,000 per show.

This show became the vanguard for a 24-hour news channel. For me it was an important milestone. I was always keen on doing something in news but had not really put much work behind the thought. From the time I had faced problems setting up Essel World, I held a grouse against the bureaucracy and its rules. They did not seem to care for the people they governed. It was tantamount to imposing a colonial rule that should have ended with the British empire. I wanted to do something to put the spotlight on such behaviour.

ON A FLIGHT from Bombay to Delhi in 1993, a gentleman walked over to me and introduced himself as Rajat Sharma. He said he was a journalist and editor with *Afternoon Despatch & Courier* newspaper in Mumbai.

'I have a programme suggestion for you that you may consider for Zee TV,' Rajat said. I was getting accustomed to people walking up to me with their ideas. In India everyone considers himself an expert. Still, I asked him what his idea was. He said, 'Zee should launch a show where political and other important personalities are grilled publicly with tough questions. These people should answer the questions on camera.' For a minute I was surprised but then realized the idea had great potential. 'Sure, we will do it,' I told Rajat. I liked the idea but needed time to think about it. A programme like this would work for Indians, who liked to question everyone. I told Rajat that I would contact him soon. I thanked him and we exchanged our phone numbers.

The idea for this show remained with me. I thought the concept needed value addition. Instead of a one-on-one question-answer session, it should be done in courtroom-style. A people's court where we would invite public figures and celebrities to answer questions about themselves. There will be political personalities, film stars, senior officials, socialites, and so on. We will put them on the mat and grill them on camera, I thought.

In a few days I called Rajat, and asked him, 'So why don't you do it for us?'

'But I don't know anything about television. I have not done any work in TV,' he said. I told him to think about it.

Then, I met my friend Gulshan Sachdeva and shared this idea with him. I asked him what kind of effort would be required to create a court-like set for this programme. Gulshan said it could be done easily as the idea looked simple. I gave Gulshan the contact number for Rajat. 'Please contact Rajat and meet him. Perhaps you can train him to be an anchor and a presenter. I don't think he knows much.'

Rajat was reluctant about this prospect and did not meet Gulshan for about two months. Finally Rajat agreed to meet me at Taj Mahal Hotel in Delhi. 'Sir, I will research the invited personality and prepare the questionnaire for the interview. But please find someone else to do the questioning for the show. I can't face the camera,' Rajat said.

Rajat seemed to have confidence in the idea but not in himself. 'We can find someone else but I want you to try once,' I told him during a meeting where Gulshan was present. 'Treat

Gulshan as your guru and start learning about TV now,' I told
Rajat. After much persuasion he agreed (though reluctantly)
and started practising with Gulshan. We created a mock set and
Gulshan began training Rajat. It took a few sessions before Rajat
could gain enough confidence to face the camera. This also
became the first show that Gulshan produced for us.

In the meantime, Kamal Morarka, the owner of *Afternoon
Despatch & Courier*, called me, 'What are you doing? This man
Rajat, I do not pay him even Rs 3,000 per month salary and you
are paying him so much to leave the paper,' he said. But I had
already asked Rajat to join on a full-time basis and offered him
about Rs 30,000 per month. I was committed and hence ignored
Kamal's comment.

We named the show *Aap ki Adalat* and launched it at the
start of 1994. Soon it became very popular. We had some very
big personalities on the show. Many were keen to be on TV but
did not know how to manage themselves on TV. Some
personalities wondered why anyone would want to be grilled
publicly. Those who were articulate and smooth talkers could
shine on the show. Rajat worked hard and did his homework
very well to ask relevant and smart questions.

With these shows my confidence also grew. I could see an
opportunity for more shows. Both *Ghoomta Aina* and *Aap Ki
Adalat* were information-based shows on a predominantly
entertainment channel. Even DD had news bulletins tossed up
between entertainment and other shows.

I tossed a new idea to Rajat. 'Why don't you also start a news
show on the channel.' This time Rajat did not hesitate. He
agreed that a daily news show would work very well. He was
enjoying the success of his show and wanted more. He agreed to
start work on it.

His excitement did not help as we realized that putting
together a news show was much tougher than doing an

interview-style show. Rajat had a print background and did not know enough about TV to launch a news bulletin. I told him to hire some people especially from DD. Rajat worked on the idea for three to four months but could not do much.

Finally Rajat confessed that he could not launch a news bulletin. So I had to step in.

I called Sam Chisholm of News Corp at Sky UK (by 1993-94 we were in partnership with News Corp) for help and advice. I told him that I would like to send a few people over for training. Rajat and about six of his team members were sent to London to be trained with Sky News.

This team learnt how to report for TV, prepare news reports, anchor shows. When they returned we put together a temporary team to produce the bulletin. Some producers and technical directors from DD would come to our studio in South Extension in New Delhi every evening to put the bulletin together. These professionals were moonlighting for Zee. Apart from the core team under Rajat, we did not hire many people. We managed with the part-time DD team.

We ran into regulatory hurdles once again. Though the broadcasting laws in India were directed towards foreign TV networks, they did not allow even domestic private networks to air news. TV news broadcast was the monopoly of government-owned DD. We had to find a way around it. News programmes were allowed to be telecast by only a select few authorized/ registered news agencies. A few foreign news agencies and ANI were the only ones allowed to uplink news content. We decided to partner ANI to circumvent the news broadcast rules. We would produce and give the news bulletin to ANI; they would uplink the bulletin to Hong Kong. ANI had the permission to uplink from India since it was an old and established news agency. But the show that they uplinked for us was on Zee TV, branded as Zee News.

We would download it in Hong Kong and play it a couple of minutes later. A delay of a few minutes allowed us to overcome the rules. We would not play it live since it would have violated the law that prevented live telecast of news by private news networks.

That's how the first private daily news bulletin of India took birth in early 1994. With this show, Zee TV became a truly hybrid channel with a mix of news and entertainment; it acquired the positioning of a terrestrial network.

Here again Rajat and his entire team (which included Uday Shankar, now CEO of Star India) worked hard. Together they made an excellent team. The half-hour bulletin aired once in 24 hours, and impressed everyone. Rajat and I developed perfect chemistry; we were constantly on the phone with each other. The conversations were both about editorial matters as well as to keep an eye on the government in Delhi. We had to be on alert to keep track of how our show was being received in official circles.

Our news show became popular and was watched in neighbouring countries like Pakistan as well. I told Rajat that we should launch a news bureau in Pakistan. I wanted to cover the country that was so important to India. I also wanted to bring both countries closer with unbiased reporting that could increase harmony. Those days Benazir Bhutto was the prime minster of Pakistan. I went and met her for her support and approval. Once both the governments agreed, we appointed journalist Nawab Kaifee our bureau chief in Pakistan. We also covered the 1997 general elections in Pakistan by sending a team from our London office. It helped that the person who was the head of Zee in the UK was a Pakistani.

At that time, Pakistan TV (run by the government) was not covering the election rallies of the opposition leaders and their parties. Both Mia Nawaz Sharif and Imran Khan were not seen on Pakistan television. Zee created a flutter by covering all the three main contenders, including Bhuttto. This election was won by Nawaz Sharif, who publicly acknowledged that Zee TV and I had played a big role in his victory. He invited me personally for his oath ceremony, though I could not attend it because of other commitments. I visited Pakistan later with my team as a state guest. Sharif told me that Zee was as important and as good as the BBC for the Indian Ocean region.

Zee was being watched by government departments in India very closely. I used to tell Rajat that our show should help our viewers. After the first month of the bulletin, we launched another programme on viewer grievances. We started a show named *Helpline*. The responsibility for this show was given to Radhika Kaul, a hardworking journalist/producer.

We would announce at the end of the show that viewers could send in their complaints to us about government departments or private companies. We would send our reporters to the departments and companies for their response. We would air the complaint as well as the response. The show helped expedite the resolution of problems and viewers were happy. Even the government departments were happy as we took a constructive approach rather than a negative stance. This programme gave me deeper satisfaction than the news bulletin. I thought I could really help people and solve their problems.

Sadly, the great run of the show was interrupted by seemingly unprofessional behaviour by Rajat and a few members of his team.

One lady executive in the team became an alternate power centre and started to dominate the rest. This was creating an unpleasant atmosphere in the newsroom. I called Rajat and said that he must remove her from the team immediately. But Rajat defended her. He said if she was asked to go, he would leave too. This surprised me, as I believed Rajat was a disciplined professional.

I did not have a choice, as I did not want to encourage

indiscipline and unprofessional behaviour. I told Rajat he could leave, too, if he refused to sack her. Rajat left the office.

But instead of leaving as one would in the normal course, he took some unprofessional steps. He phoned everybody in the team and asked them to quit as well. About 60 per cent of the team decided to leave with him, the same day. By afternoon, much of the team had left. Rajat announced that he would make sure that the news bulletin would not go on air that day and even later. He even called and asked ANI not to dispatch the bulletin. All this within a day.

Jawahar and I wondered why Rajat was behaving like this. Until 24 hours ago Rajat would not tire of telling everyone how Subhash Chandra had made him a hero out of nothing. In public he had said that he would pledge his life for me. 'I would never forget that you are like my godfather,' he would tell me publicly. Jawahar was with me in the Delhi news studios. We had rushed to do some firefighting. We asked the DD team to come in early and resume work with the remaining team. The DD producers were very helpful. They promised to ensure that the bulletin aired on time. The remaining 30-40 per cent people left in our team worked hard and the operations went smoothly that day. Though it was a day of crisis, we survived. After that day we did not face any such problems.

In fact, some of the people who Rajat had taken with him, returned to our fold.

I continued to pursue the logical extension of the daily news show—a 24-hour news channel. The birth of Zee News was inspired by CNN and BBC. Why couldn't India have a news channel? I don't think we really made a business plan for it. We needed a news channel. The half-hour show had given me the confidence for it.

Soon after, in 1995-96, we launched India's first privately-owned 24-hour news channel, called Zee News. Due to the

Rajat incident, the news team had become divided and political. There were many factions. Even after he had left and our 24x7 channel was launched, these factions continued and the infighting culture remained. Jawahar took charge of the channel to make it efficient.

The impact of the negative work culture hurt us in the long run. When Aaj Tak launched in 2001, we lost the number one position very soon. Our team had become complacent because there was no competition to Zee News until Aaj Tak was launched. Aaj Tak was first a show on DD produced by India Today Group's promoter Aroon Purie. Later it was launched as a 24-hour news channel. Before Aaj Tak was launched, Star TV had news bulletins produced by NDTV's Dr Prannoy Roy.

There was a sudden explosion of news channels between 2001 and 2006. During this period Zee News had to work hard to maintain its position.

So far we have not been able to regain the number one position. Star News launched with NDTV and then later on their own. Meanwhile, we also launched a business channel, Zee Business, and a few other regional news networks. In August 2014, we had 160 million viewers of our news network in India (as per TAM data).

India has witnessed the birth of hundreds of news channels in various Indian languages. Some of them are owned by criminals and history sheeters. The respective governments have closed their eyes, taking shelter under the guise of freedom of speech. I predict that history will not forgive respective governments and their information and broadcasting ministers for not keeping a sharp eye on the quality of people who launched these channels.

Issuing a licence for a news TV channel is as important for society as issuing a licence for a bank. While there are strict restrictions on bank ownership, there are almost none for news channels. A news channel run by an irresponsible person can severely damage our society and country. It may be possible that today the well-known don and enemy of India, Dawood Ibrahim, owns a news channel through front companies. The government must wake up and create strict rules for news channel ownership. The security clearance must be similar to the 'fit and proper' criteria that the Reserve Bank of India applies for banks. This is critical for a country the size and diversity of India.

MY PROFILE AND reputation in the government changed after the news bulletin started airing on Zee TV. I was received very differently by the system. I said to myself, 'Subhash, this box has given you a different profile. You should not think that you have created the Zee network. Rather you should think that the Zee network has created a new you.' It was important to remain grounded despite the fame and attention that media brought.

We did manage some early innovation that set the stage for channels that were launched later. We started a weekly programme called *Hamarey PMji*. A camera van would go to different places to allow citizens to record their message for the prime minister in 2 minutes. Each person could directly address the PM through our camera. Those days P.V. Narasimha Rao was the prime minister. We would collect the relevant questions for the PM and send them to his office. The PMO would respond to some of these. The PMO was very happy with this show as we would air both the question from the public and the response from the PM's office.

During the same period, general elections were announced. We decided to launch a programme in which we covered all the 540 parliamentary constituencies with our team of TV journalists. About thirty to forty production units worked on this show, which was a run-up to the selection of candidates for these elections. For each constituency, we did a profile of the sitting member of Parliament, his/her work and his/her chances of winning in case he/she again contested. We also profiled the chances of an alternative candidate from the same party and, of course, other parties.

We would sum up and say which party or person stood the best chance of winning. We did not have any agenda nor were we particularly in favour of or against any party. But as the show progressed, certain trends became apparent to us. It began to appear that many sitting Congress MPs as well as MPs from other parties would lose their seats.

We mentioned the likely winners in our reports. But it was a mixed bag and the majority were not from the Congress. This ruffled feathers in the Prime Minister's Office. Narasimha Rao would himself tell me, once the dust had settled, that Matang Singh, his minister in the PMO, had told him that Zee was running a campaign against the Congress party. Instead of analysing the show and its results, Matang Singh seemed to have convinced Rao of the need to either stop the Zee network or bring it to its knees. He wanted Zee to favour the ruling party. Singh apparently got the *Asian Age* to write front-page stories about foreign exchange violations by me and my group.

I was in London those days. I had to remain abroad for some more weeks that year to maintain my NRI status. These news reports created a lot of tension for my brothers and many senior Zee executives. Around the same time, our entertainment channel committed a blunder during the New Year's Eve special programme, by airing a spoof of Prime Minister Rao and his so-called guru Chandraswami.

After it aired Jawahar called me in London. 'The government feels that due to our past relations with the Gandhi family, Zee is against Narasimha Rao's government.'

We had earlier decided that we would not do any programme that would insult or denigrate key personalities like the president, prime minister, governors, Supreme Court or high court judges. Since our channel was seen in many countries, we did not want our leaders to be ridiculed. Despite this rule, the spoof on Rao seemed to confirm to the government that the Zee Group was against it. Jawahar sent me the tape of the skit and I understood the impact it would have on Zee. I called Prime Minister Rao from London. 'Sir, there is a big blunder that has occurred on Zee TV,' I said.

'What happened? I don't know anything,' Rao said.

'Sir, there was a spoof on you that went on air. It was done by a young producer who we have removed. I don't know what else to do now apart from sacking the producer. The arrow has left the bow but I would like to apologize for it,' I said.

'No no, Subhashji, it's OK. You have explained the skit to me. And we should laugh it out. However, we Indians don't know how to laugh at ourselves,' Rao said, and dismissed the issue. This was not the end of the issue, though.

Within a week of the airing of the spoof, we were raided by Enforcement Directorate officials who accused us of foreign exchange violations.

My brothers were in a panic and informed me in London about the raids. Jawahar asked me to stay on in London. He feared ED would arrest me if I returned. But I did not care. I returned within a day of the raids and met the ED officials.

After the elections were done and dusted, Rao would himself admit to me that this raid was in retaliation to the nature of our programming. At the time, despite my apology, the PMO was attacking me. I had no choice but to fight back. I called the producer of *Hamarey PMji*, Seema Murlidhara. I told her not to edit out difficult questions put to the PM and that if the PMO did not answer these, the programme should be critical of the prime minister and his government.

We played up questions from viewers that the PM had no answer to. We ran these questions non-stop, and it became embarrassing for the government and the prime minister. Within ten days, we got a call from the PMO.

A. Prasad, an IAS officer from the Andhra cadre who was OSD to Rao, called me for a meeting. I met him at his South Block office. 'I hope the PM is happy with Zee for airing the show,' I said.

Instead of talking about the show, he asked me why the Enforcement Directorate had raided me. 'You are our friend, and I will ask the revenue secretary to stop it.'

I responded by saying, 'Prasad sahib, the revenue officers are doing their job and let it be.' He did not know where to look and to be fair to him, perhaps he didn't know that the raids had happened with the knowledge and instructions of the PMO.

Now he came to the point and literally begged me to stop the show. I told him that the show had nothing to do with the raids by the Enforcement Directorate and that they were just doing their job. The matter ended there. The show continued, and so did the investigations against us.

In the ensuing elections, the Congress lost. There were many reasons for the loss, including the demolition of the Babri Masjid. Our show had captured the election trends accurately. Rao met me after the elections and almost cried: 'I wish I had paid attention to the reports you put out on your channel.'

'If you had believed us, you may have returned as PM,' I told him. Rao agreed. I told him to look at the tapes. Almost 95 per cent of our predictions were true.

Rao had been misled into believing that Zee was against him. And therefore, he did not take our reports seriously. He paid a heavy price for it.

I STAYED IN Hong Kong from 1993 to 1996. I would shuttle between London, Hong Kong and India. On most days I would work 14-15 hours a day.

From living in a dal mill to a big house in west Delhi and from there to live in a house in Mumbai had been big steps for me. But this jet-set life was very different. In Hong Kong I stayed in a room within the office. I was on phone all the time. I would be in touch with Karuna, head of programming, and Digvijay, who was business head. I managed to resolve issues, take decisions and push for growth without physical meetings. I would meet them twice or thrice a month.

We decided to list Zee Telefilms on the stock markets. I chose to raise resources in a professional way. In the process, I had to explain to the market that Zee Telefilms was a content producer, not a broadcaster, that it was procuring programmes and exporting to Asia Today, that it made 10 per cent on programming cost and 15 per cent on advertising revenue. The IPO was oversubscribed by 60 times. The Rs 10 share of Zee Telefilms opened at Rs 60 on the day of listing.

18

A ROCKY PARTNERSHIP

*Dealing with Mr Murdoch—ally and
rival both*

AT THE START of 1994, Li Ka-shing and his son Richard sold
Star TV to Rupert Murdoch's News Corporation. I learnt that a
senior partner of Goldman Sachs had brokered the deal between
Murdoch and Li during Christmas and New Year holidays on a
luxury cruise boat. Murdoch was told that over 20 million
households (and counting) in Asia were watching the Star TV
network. Murdoch was paying $2 billion to buy Star TV.

When News Corp's representative Sam Chisholm took over
Star, he soon realized that of the total 20 million households,
the authentic number was only 15 million. Moreover, 8 million
out of the 15 million did not really belong to Star but to Zee TV
in India, a company in which Star had no ownership. Star made
about $4 million profit per year from Zee by leasing its
transponder on AsiaSat. As mentioned earlier, Star was renting
the transponder for $1 million from the owners of the satellite
but getting $5 million from Zee.

After getting into these details, News Corp decided to send
one of its executives, Andrew Carnegie, to India. We knew
about it from the grapevine but he did not meet me for a few
weeks after coming to India.

Star TV had a representative in India, Siddharth Ray, but when News Corp took charge they also appointed their man, Iqbal Malhotra (who reportedly happened to be an acquaintance or relative of a senior executive at News Corp). The original Star TV representative was sidelined. Rumours started that Zee TV would close down as Star TV would launch its own Hindi channel, that Murdoch wanted to cancel the transponder deal with me.

To prevent a disaster, I got in touch with our partners, the private equity investors (a fund owned by Sir James Goldsmith and Kerry Packer) in Hong Kong, and said that we should start planning for such an eventuality. I was not sure if Star would take such a step but I did not want to be caught on the wrong foot. If News Corp decided to part ways, I would have to be prepared. I told my investors that we should look for a Russian satellite and rent a transponder on it. We had to move fast as we did not have much time. The transponder deal with Star could be cancelled by either party by giving only a six-month notice.

The investors were worried about the impact on viewers. All the satellite dishes in India were facing AsiaSat. And migrating to another satellite meant shifting the alignment of all the dishes. I was not too worried as I thought that Zee had enough influence among viewers to force cable operators to change the direction of their satellite dishes. I believed that if Zee moved, dishes would move with it and Star TV's viewership would be affected.

After a few days I got a call from Andrew Carnegie of News Corp and we met at The Oberoi hotel in Mumbai. He took charge of the meeting and began the conversation as if he owned me and my company. He said Murdoch was not happy being only a lessor of the transponder but wanted to start his own channel. I said, 'You cannot start another competing channel as per our agreement.'

He agreed but said that he could cancel the transponder lease and the agreement with Star.

I said, 'Yes, you can cancel the contract.'

And before he could say another word, I took a document from my jacket and handed it over to him. 'This is my termination notice. I don't know whether you will cancel the agreement or not, but I am giving you six months' notice to end the agreement.'

Andrew was shocked. He did not know what to do or say. I had come prepared for this conversation. I wanted the upper hand in the situation and managed to get it. Andrew was expecting a meek Indian partner who would be scared of ending the transponder agreement. Instead, he faced a belligerent one. Before he could react, I ended the meeting and left.

This was a calculated gamble on my part. My investors were not supporting the proposed shift to a Russian satellite. My strategy was to precipitate the matter for a quick resolution rather than fretting for days. I was trying to force News Corp's hand. If News Corp did not cancel the deal, it was okay. But if it did, I would then be able to persuade my investors to look for an alternative immediately. The earlier we chose our path the better it would be. Soon, Murdoch's right hand man, Sam Chisholm, called. He was based in London and Andrew reported to him.

Sam said, 'Can you come to Hong Kong to meet me? We can have a drink together.' I agreed, though I did not know what to expect of this meeting. I gathered all my investors and brought them up-to-date on all issues. They felt that News Corp would want to buy me out, but that if I did not sell, they might keep me as their partner for some time. They advised me to appoint an advisory agency for this meeting. We appointed a boutique merger and acquisition advisor in media industry, called Communication Equity Associates (CEA).

I took a CEA representative along with me for negotiations

and to navigate legal issues with Sam. We met Sam and explained that we were not desperate to remain with Star. If Star wanted to stay with Zee then it would have to come up with a good deal. More than half the subscribers of Star TV were with Zee. If we had moved out of the platform, the Star network would have lost significantly.

The negotiations took about two to three weeks. CEA and my equity partners were giving me advice and inputs. Finally, News Corp decided to buy a stake in Zee. They did not want Zee to be so independent that it could walk out any time. Also, to maintain its own business valuation, it was important for News Corp to take a stake in the company that supplied a big chunk of the consumers in Asia. This suited me as well. Instead of looking for a Russian alternative, I was getting a partnership with a global broadcaster. News Corp took me seriously and treated me like an equal. Andrew's original attitude had disappeared. The News Corp team no longer thought of me as a small-time operator.

We decided to become partners. News Corp bought 49.9 per cent of Asia Today for about $60 million.

This transaction was done in sequence, yet in parallel. First I bought out the private equity investors. Initially, the investors had given me $8.6 million for their 49.9 per cent stake. Now I bought it back from them for $30 million. Next I sold this entire stake further to News Corp for $60 million. Everyone was a winner this way. The patience of the investors had paid off. They made more than three times' profit on their investment, in less than two years.

News Corp reduced the rent for the transponder from $5 million to the prevailing market rate of $1.2 million per year. Richard Li's original figure was now in play.

We entered into a shareholders' agreement under which it was agreed that the joint venture would undertake programming in all Indian languages, while Star would remain only in English. This was to ensure that there was no competition between the partners.

News Corp also became investors in our cable company, Siti Cable. They invested 49 per cent in the cable company. This was the maximum equity limit allowed under law. They wanted to partner with me in everything. They were also keen to go deeper into the broadcast industry with my help. Between 1994 and 1996 the partnership progressed very well.

I met Rupert Murdoch for the first time in late 1994 in Los Angeles at his residence, and we agreed to enhance the partnership by launching an additional TV channel, Zee Cinema, together.

It was a simple meeting and we exchanged pleasantries. My main contact was Sam Chisholm.

I had several one-to-one meetings with Murdoch. Initially, between 1994 and 1996, we would meet as partners and plan our next moves together. For the Siti Cable deal, we met in Los Angeles. Apart from this he asked my opinion on what could be done in the Indian market.

Many people had warned me against doing business with Murdoch. Globally, his image was of someone who did not believe in partnerships. I tried to protect myself in every way possible in the shareholders' agreement.

It turned out that News Corp had been smart enough to buy my stake at a price much cheaper than its market value. I discovered this when the figures of News Corp's buyout of Star were revealed. News Corp had paid Li Ka-shing and his son Richard close to $2 billion for a market of only 15 million homes. Out of these 8 million were Zee homes. By that calculation, my valuation should have been about $1 billion. And therefore, the value of the Asia Today stake should have been $500 million. Instead, News Corp had bought my stake for just $60 million, though my sale had created waves in India. Star had already launched Star Plus, Prime Sports, MTV and Star Movies in India before News Corp took over. After the takeover, they continued to push for more global channels. The entry of the Star brand in India was seen as a sign of a new emerging India. These were the early years of reforms. India had opened up to foreign investors and had ended industrial licensing.

In 1995, when Murdoch came to India, everyone wanted to meet him. And he was keen to meet the leaders of a new and emerging market. We arranged meetings with the who's who of India. These included the president, the prime minister, Cabinet ministers, including the information and broadcasting minister. Murdoch also met top corporate leaders and attended dinner parties arranged by us for corporate India and the movie industry. Rathikant Basu, who headed Doordarshan, resigned from the Indian Administrative Service and joined News Corp.

That's when the relationship between News Corp and Zee started going downhill. Basu said Star should do shows in Hindi; he launched these shows without any discussion with us. Basu misinterpreted the shareholders' agreement and probably misled News Corp as to how much Hindi programming was allowed. Our deal was clear: Zee would handle all Hindi channels and programming and Star would focus on English shows and channels. Neither would venture into the other's market.

WE HAD EXPANDED our operations in international markets. Zee was being beamed in continental Europe, the UK and the US. This international operation was outside the Zee Telefilms or the News Corp partnership. Our partnership was restricted to Indian markets. They were not interested in Hindi programming globally.

Then Murdoch's representative Malhotra started bad-mouthing us publicly; he was also poaching people from our company. In a clear violation of our agreement, Star TV started programming in Hindi. This created tension in our relationship. I was told that they were using all methods crossing boundaries of gathering business intelligence. These were tough pressures. I kept requesting Gary Davey, the CEO of Star TV Asia, to stop violating the shareholders' agreement. However, there was no resolution of the problem. We had no choice but to issue legal notices to them, referring to various clauses of the agreement they were violating.

The key reason for this behaviour, it seemed to me, was that News Corp and Murdoch were not used to dealing with a partner who was in the driver's seat. Even worse for them, I think, was that we dared to question News Corp.

The management control of Asia Today was with me and News Corp found it tough to have their way with me. Soon, they began to find faults and loopholes in our joint venture. They would send their auditors to check if we had made any mistakes in accounts or tried to mishandle funds. They were breathing down our neck all the time. But they could not find any loopholes or misdemeanours in our working.

Despite this, I was keen to expand into related categories of the broadcast sector with Murdoch. The nature of the media business is such that there are no permanent friends or enemies, and News Corp and Zee are still partners today in a channel distribution joint venture named Media Pro. Thus there was no lasting enmity with News Corp. When Murdoch came to India in 1996 on his second visit, I made a suggestion. I said, 'Let's have a joint venture and start direct-to-home distribution platform. There is no regulation/licensing requirement at the moment and we can quietly start the operations.'

This, however, upset Murdoch. 'What do you have to do with DTH, that's my domain,' he almost scolded me. I kept quiet. I did not agree with his view that DTH was his monopoly in the world. He might have control of it in the UK market, but he could not claim rights over India. His refusal did not deter me from my plan to get into the DTH business. I began looking for other options and possible new partners for starting DTH in India.

I went to Malaysia in 1997 and met Ananda Krishnan, promoter of Astro Satellite. Krishnan's parents were Tamils from Sri Lanka. As such, he could also claim to be a person of Indian origin. He had launched a satellite and was in the process of launching DTH operations in East Asia. He had a monopoly licence for Malaysia. During a dinner hosted by him for me in Kuala Lumpur, I asked him if we could partner him to launch DTH in India. He refused. 'No, I can't do it with you. I will want to do it with the Government of India or on my own. I have a lot of relationships in India and I can work alone.' He already had three or four transponders with an Indian footprint.

Krishnan then apparently used his influence with the then Indian prime minister, P.V. Narasimha Rao. He worked through the Malaysian government to meet Rao during the latter's visit to Kuala Lumpur. Krishnan also got in touch with the Indian Space Research Organisation (ISRO) as it was a key body that controlled satellite policy in India. Krishnan managed to sign an MoU with the Government of India, through ISRO.

I had thus been rebuffed by two players who could have been my DTH partners. I could not think of anyone else to partner with, and I did not want to waste my time looking for others. But I was not going to give up without a fight. The prospect of foreign companies being able to launch DTH without any restrictions, while domestic players were made to stay out, seemed absurd to me. Meanwhile, Murdoch also got busy trying to launch the DTH operations in India. He hired Urmila Gupta, an officer of the Indian Information Service, to start the DTH service, under Rathikant Basu.

We decided to take on both Krishnan and Murdoch. My team and I began meeting policy-makers and ministry officials to educate them about DTH operations. I told them that no country allowed foreign ownership beyond a minority stake in such ventures. The reason was that there was no filter between DTH broadcasts and consumers. Unlike cable, DTH could not be controlled by the government. The technology could create targeted communication with a person or a group of persons. This had scope for misuse.

By now I.K. Gujral had taken over as prime minister. All efforts were made by the global players to convince the Gujral government about the need for DTH in India. Some domestic TV industry players and modern-day turncoats also began lobbying for foreign DTH players in India. These were similar to Indian rulers who had helped the British establish their rule in India. I believe that Gujral was persuaded by a prominent TV production house owner, who later became a broadcaster, to allow Murdoch to launch DTH operations. This producer was already selling content to Star channels.

While Gujral began to be convinced about foreign DTH, the information and broadcasting minister, Jaipal Reddy, was not so sure. He checked and found out that the Telegraph Act did not allow such services. The law was not clear on DTH services because it was a new technology.

He consulted the then Cabinet secretary, T.S.R. Subramaniam, and decided to issue a notification clarifying the issue and to remove any doubt about the status of DTH. The notification disallowed DTH services without proper permission or licence. The issuance of licence was not possible as the legislation on DTH was being considered by the government. Until the bill was prepared and enacted into law by Parliament, DTH would not be allowed into India. The foreign players went to Gujral and informed him about the impending notification. Gujral was livid with his Cabinet secretary. He pulled him up for trying to issue such a notification without checking with him. 'Are you the prime minister or I am? What are you doing?' he shouted at the Cabinet secretary.

Subramaniam replied politely, 'Sir you are the prime minister, but there is a Cabinet resolution prohibiting DTH that is still valid. If you want to allow it, it can only be approved by the Cabinet. Hence we will have to call a Cabinet meeting and ask it to reverse the existing resolution by passing a new one.'

This stumped Gujral. He did not have the courage now to reverse it. He was the prime minister of a minority United Front government that depended on outside support from the Congress. The Cabinet had members of different parties, including the communists. If he tried to reverse the earlier Cabinet decision, he would be accused of pandering to foreign interests. As a result, the notification was issued and it effectively stopped the foreign companies in their tracks. They could not enter DTH until the legal framework was ready.

This stopped Murdoch from starting his DTH operations just one day before his launch. Of course he got even with us later. When we finally launched our own DTH business, Dish TV, he did not provide the Star bouquet for two to three years until we got a Supreme Court order in our favour.

I discovered later that other players were also busy stalling foreign DTH. Shiva Shankaran, a serial entrepreneur from Chennai, had tried to sway the bureaucracy against FDI in DTH, and rightly so.

As a result of all this, even the MoU between Ananda Krishnan and ISRO could not be converted into a DTH operation in India.

IT WAS NOT just DTH. Foreign players were beginning to realize the importance of the Indian market and had started making attempts to take over the Indian media industry.

News Corp and other foreign groups began lobbying for an increase in foreign holdings in media companies. In television, cable and programme production, 26 per cent to 49 per cent foreign holding was on the anvil. But the foreign companies wanted 74 per cent to 100 per cent.

All the excitement and success of Zee during 1994-95 had drawn the attention of the government. The first Indian private channel was a commercial success, but was still illegal. There was obviously a need to draft a comprehensive new broadcasting law.

A parliamentary committee was constituted in 1994-95 to formulate this new law. By then I had also met many political leaders in Delhi to explain how satellite broadcasting worked. At that time the entire media and entertainment industry, including TV, movies and music, was about $300 million (the Indian rupee was 30 to the US dollar). Today, this industry would be worth $16 billion (see Appendices). I made presentations in Parliament to many members and government officials on the future of broadcasting. I would tell them that the $300 million entertainment industry would grow to $3 billion by 2000. I wanted to improve the understanding of MPs so that they could contribute to creating a modern and forward-looking legal framework for the broadcasting sector. After one such meeting, the owner of the Jagran Group, the late Narendra Mohan, who was also a BJP MP, came to me to find out more. He was incredulous and thought that I was exaggerating. But I explained how this could happen. A lot of people those days did not realize the potential of the broadcasting sector. While I tried to create a more amenable atmosphere for TV industry in India, the environment was changing globally.

We met all the stakeholders in India—the political class, the executive, the corporates and civil society. The idea was to make them aware of foreign ownership restrictions in other countries. We were giving special attention to the visual medium since it could reach even the illiterate population. The newspapers could be read only by a small percentage of the population that was literate.

By 1998-99, News Corp had violated the shareholders' agreement in terms of Hindi programming and seemed to want to take Zee over. I had to counter it by reaching out to key institutions and people—leaders in corporate India, policy makers, media owners and influential people—and telling them what was happening.

Most people we met were sympathetic but were not sure how to help us. The late Aditya Birla of the Birla Group became a staunch supporter of Indian media. He was ready to support us financially. He told me, 'Subhash, do not sell for want of finance. You have a blank cheque-book at your disposal from me. Media should not go in the hands of outsiders.'

We also approached the Rashtriya Swayamsevak Sangh (RSS), an organization known for their nationalistic agenda. We asked for their support against the enhancement of foreign investment in the media sector. By now an RSS-supported BJP government was in power under the leadership of Atal Bihari Vajpayee. The Left parties also supported us when we explained the situation to them.

As an individual, I was not against FDI but I had seen how the laws were being used by anti-national activists. To my mind, instead of respected foreign media companies owning the Indian media, the structures could allow anti-national individuals to own the media. There could be other organizations, possibly criminal, who could exploit the weak regulations to bypass the law. Even now there are so many news networks whose real ownership is hidden and unknown. At the early stage of an immature market, it would have been terrible to have unknown entities controlling broadcast sector. Even today, ownership is not entirely transparent.

I met RSS chief Rajju Bhaiya in Nagpur. He did not understand much and asked me to meet the second-in-command, K.S. Sudharshan. In a meeting lasting four to five hours at RSS's Jhandewalan office in Delhi, I explained the issue to him. He took notes of everything. I think they convinced the Vajpayee government to limit FDI in media.

DESPITE RATHIKANT BASU'S best efforts, Star continued to lose money in India. So Murdoch decided that it was time to exit India. He felt that he could not run the business profitably here. Inspired by Zee's success, even *Business India* magazine had launched a channel, BI TV. However, it was losing money. UK's Pearson Group and the *Hindustan Times* Group had partnered to launch Home TV. But that too lost money and folded up. The failure of these partnerships appeared to convince Murdoch that India was a tough market to crack and make profits in.

He thought it would be better for his partner Zee to run the network profitably. This would also end the acrimony and competition between Star and Zee. He realized that Zee had somehow found the formula to be successful. Murdoch admired me for my success but disapproved of me for challenging his organization. Also, I feel, he could not understand why Star was not as successful as Zee despite the growing market. What was I doing that his team could not do? Some industry insiders feel that even after twenty-five years of being in India, the Star network has not made net profits. In 2015, too, the profits of its entertainment channels will be smaller than the losses in the sports category. But all this is speculation as their numbers are not in public domain.

On the other hand, Zee has been giving 30 per cent return to shareholders year on year. There were two years, though, when we did badly but that was because of factors beyond our control.

The first was in 2000 when the markets crashed. And the second was in 2008 during the global financial meltdown. Most other years, our profit margins have remained steady at around 30 per cent.

To finalize the terms of his exit from India, I met Murdoch with Basu and my friend and advisor T.N.V. Iyer, and a representative from Goldman Sachs who was managing their exit. We met in London at Murdoch's office.

The plan was to merge Star's operations in India into Zee. Murdoch would get about 22-26 per cent (minority) shares in the merged entity in return. Goldman Sachs hammered out the documents within two days in London. This was the only time that Murdoch and I were involved directly in negotiations. All other deals were done by his executives and me. But for this one we sat across each other. He came to my London home for dinner and had also invited me to his home in London. I had a lot of respect for him, and as a result agreed to give him a higher stake than business sense demanded. Murdoch had that kind of presence. Ideally, he should not have got more than 15-17 per cent, according to our advisors.

We started work on the regulatory process for Star's operations to be merged with Zee. Basu came and congratulated me. He said this deal would be good for the Zee Group. Even though he had started the problems between Star and Zee, we had kept a role for Basu in the post-merger scenario. I thought he should be the chairman and hold a titular and ceremonial role to ensure and maintain a smooth merger.

The Vajpayee government was in power while we were negotiating Star's exit. Somehow, the government came to know what was happening. In one section of the government, panic bells started to ring. Some people in the government and the BJP felt that if Star exited/merged with us, the Zee Group and I would become too powerful. Our detractors were prominent business group(s). They were trying to persuade the government to stop the deal between Zee and News Corp, and they succeeded in their aim.

I got a call from Murdoch, who said, 'You don't seem to have good relations with the Indian government.' He then told me that Sushma Swaraj had met him in New York and told him that the Indian government was not in favour of the Star-Zee deal. She had apparently promised support to News Corp for their businesses in India.

I was informed by a credible friend in Vajpayee government that while the official reason for Swaraj's going to New York was to attend the UN general assembly, meeting Murdoch was also an important objective of the trip.

No surprise then that Star TV decided to oblige the Vajpayee government and stayed on in India. I don't know why Swaraj was made to do this. Though I knew who was plotting against me, I could not understand why the Vajpayee government felt threatened from me.

The irony is that the Congress party felt I had a soft corner for the BJP, while some leaders in the BJP felt the reverse. Despite my friendship with the Gandhis (I had disclosed to Rajiv that I was a *swayamsevak*; he had just laughed it away), the Congress saw me as a BJP sympathizer. On the other hand, Vajpayee and his close circle knew about my background in the RSS, but this group was seemingly against the RSS. I tried to ask the government why the deal and the exit of Star was being stopped but did not get any answer.

Despite Murdoch's decision to stay, one part of our deal went ahead. We bought their stake in the Zee group in both content as well as the distribution company for a hefty sum. For Zee and Siticable it was $296.51 million. The deal was part cash, part shares of Zee. This was the time when the new media and dotcoms were close to peaking. Zee's share price was increasing daily. The news that we had bought back the News Corp stake further fuelled the share prices. The newspapers were tracking the wealth of new economy entrepreneurs like Narayana Murthy of Infosys, Azim Premji of Wipro, Mahendra Nahata of HFCL and me as a rising media moghul.

Stockbroker Ketan Parekh started investing heavily in ten stocks, including Zee. These ten stocks were called K10 by the market and the media. All of them were rising on the bourses. One day, when I was on a morning flight to Bangalore, I saw my name on the front page of a newspaper; it said 'Subhash Chandra, the richest Indian.' This was when Zee's Re 1 share was quoted at Rs 1,500.

Murdoch returned to India and sacked Basu. He appointed Peter Mukerjea, who was in charge of ad sales in Star, as the head. Under the new leadership, Star launched *Kaun Banega Crorepati* and hit the road to success. Initially, they wanted to launch with Rs 1 lakh or Rs 10 lakhs as maximum reward. But Murdoch was not convinced. He asked whether a house could be bought in India for Rs 10 lakhs. His team said no. He asked what the next denomination was. He was told it was Rs 1 crore. So he cancelled the original figure and said it had to be Rs 1 crore.

Star also took a big decision in making Amitabh Bachchan the host of the show. By then he was seen as a has-been superstar. But the decision to make him the KBC host turned out to be great. Both Star TV's and Bachchan's fortunes changed thanks to the success of KBC.

KBC could have been hosted on the Zee network, but for a big error of judgment. The KBC format had been offered to us first. But the then CEO of Zee, R.K. Singh, turned it down. I got to know about the offer much later.

19

SETTING MY HOUSE IN ORDER

A family division...and a Tehelka *sting*

SOME TIME IN 1997, my father called me. He was upset and said that I was ill-treating my brothers. He told me to give them whatever I wanted and officially separate them from my business. 'Give them their share and let them be, you have been an irresponsible elder brother,' he said.

This came as a shock to me. It was only after my father's call that I realized that my brothers were upset. It was then that I discovered what the media business was all about and that people like Jindal had been creating a wedge between us brothers. My brothers had deep respect for me, and so did not ever question me. Many people, including Jindal, used my name to take steps that they wanted. I remained unaware of how they were using my name to belittle my brothers. It was not as if my brothers and I agreed on everything but we trusted and respected each other.

Some wise people have said that families that own and run a media house can't stay together. This almost happened with us.

I moved quickly on the matter and met my brothers and our father. I explained that I had not pushed them away from the business. I told them I was not greedy about the business we

191

had created, that in the media business, it's important to have one leader. Many leaders would pull the company in different directions and hurt the business. I offered to my brothers that any of them could take charge of the business. I would relinquish control, if it helped us stay together. But if I was in charge then the business should run my way.

My brothers were not convinced by my apologies. They were angry and hurt. They had been suffering insults for many months. They wanted to separate. I was distraught and I wept.

I couldn't separate from my brothers because of silly misunderstandings. We had built our business together. We had supported each other and done everything required to ensure that we succeeded. I could not let years of our effort collapse.

But I realized that it was important for me to create space for everyone and establish well-defined rules that would enable us to work together without future misunderstandings. From a small business, we had created a large group. It was important for everyone to have the space to grow.

I drafted an agreement and sent it to all three brothers. The idea was to have clear and common ownership of the group. I proposed that the four of us and our father would own 20 per cent each of all our businesses. All of us would be equal partners. That was the spirit always, not only in our family but also in most joint families in India.

So far we did not have any written agreement on the ownership of the group companies. We had all assumed that we were equal partners. Shares of different businesses could exist in anyone's name. We did not really care who owned what on paper.

At that time, in 1997, the main companies of the group were Zee Telefilms, Essel World and Essel Packaging. There were several smaller businesses as well.

But my brothers and father did not accept my proposal. They said I had built the businesses so I should run them. In most families, when business is divided, everyone wants more. But it was the opposite in our family. My brothers felt that my ownership share should be higher than theirs. Also, my father said that he did not want to own any shares. We were not able to agree to this arrangement, and sought help.

We requested a family friend and former banker, A.C. Saha, to speak to everyone and find a solution. He met all the brothers and discussed the options with them. Finally, an agreement was reached. My brothers wanted me to hold 40 per cent while each of them would hold 20 per cent. My father had relinquished his shareholding. My brothers also supported the idea. I did not want to hold more than others, but finally agreed to this arrangement.

We started work on implementing this in each of the companies. But it was slow and nobody was really in a hurry. That meeting resolved most of the issues between us. My brothers realized that I did not want to hurt them and that I did not want to exclude them from the business.

But after this family meeting I decided to ask Jindal to leave. I brought in R.K. Singh to replace him. And then I asked Jawahar to take charge of Zee News. Later, Laxmi took charge of it as Jawahar had to handle Siti Cable and DTH operations of Dish TV.

THREE CEOs FOLLOWED Vijay Jindal, but none could bring the flagship Zee TV to number one slot. Zee TV was number two when Jindal left but slipped to number three afterwards. Despite this, I kept adding new channels to the bouquet. And this allowed the company's profits and revenues to grow. As stated earlier, in the twenty years of the Zee network's existence so far, the results have been flat for only two years. And in those years the market conditions were very poor. The average growth of our media and entertainment business for over twenty years was in excess of 30 per cent.

From Zee's launch in 1992-93 itself, we had started morning-band programming on Zee TV. In this one hour we would air discourses by different religious speakers. The show was called *Jagran*. I discovered that some members of the Zee team were being bribed by these gurus to air their discourses. The following of these gurus was multiplying after they appeared on Zee. These shows were free promotion for them, whereas our intent was to help our viewers. So many of the new gurus were desperate to appear on Zee. And they were willing to pay the channel or bribe my team for it. I was angry and I cancelled their appearance and aired the discourses of only those I thought were clean.

Later, we launched a 24-hour channel on spiritual teaching called Zee Jagran. And then we also launched Zee Salaam covering Islamic teachings.

Siti Cable was moderately successful. Zee Cinema, Zee Marathi, Zee Bangla, and ETC were also making money. Even the international feeds were making money. De-risking the business—by launching more channels and earning revenues from global broadcasts—helped us survive the lower profits of the flagship channel Zee TV. We lost many good people because the CEOs were not able to deliver and were not able to nurture the talent in the company. There is a popular belief in management: people join a company, but leave their managers. I tend to agree. Good leaders are rare. Even though we were getting smart professionals, they were leaving since their bosses were not inspiring enough.

Our experience with the next CEO, Pradeep Guha, was not too good either. Towards the end of fiscal 2006-07, we finalized the budget for 2007-08. We took approval of the board of directors. The Zee network's estimated profit for the year was about Rs 600 crores. One day in June 2007, Pradeep called me on the intercom and sought a meeting for a presentation. He asked for one hour of uninterrupted time.

We met for about two hours. He told me about the competitive landscape and new launches. Peter Mukerjea, the former CEO of Star, was launching 9X, a Hindi general entertainment channel. It was backed by Singapore government's fund Temasek and a large Indian corporate. Viacom of USA was coming in with TV18 and also launching a Hindi entertainment channel, later branded as Colors. They had poached on our programming resources. 'The sponsors of these new channels have deep pockets,' he said. Meanwhile, Star TV was becoming even more aggressive. After outlining these details, Pradeep came to a surprising conclusion. He said that our flagship Zee TV would come under pressure and lose ratings even further.

He said, 'I think we should forget about profit for this year. We will have to invest a lot of money in programming and other critical activities.'

I was shocked. 'Pradeep, what will I tell my shareholders? We have projected a profit of Rs 600 crores and that is public information. Now how can I go to them and say that there will not be any profit. We will be finished. The shareholders and analysts in capital market will give up on us.'

Pradeep had a counter argument. He wanted Zee to almost treble Zee's programming spend to counter the challenge. But if Zee were to increase its spending on programming, there would be little profit left. He told me that all the new channels were spending hundreds of crores on new programmes, distribution and marketing. Colors, for instance, was planning to spend crores on reality shows alone.

I was not convinced of the need to increase our spending. 'If the new players are spending so much, let them do it. Our plans should not be affected. I do not want to get into the rat race with the new players.'

If I was given an assurance that by spending more money our rating would go up, I would be willing to spend. But Pradeep did not promise me higher ratings in return for the extra spending. I said, 'Pradeep, let me think about it overnight. We will discuss tomorrow and take a final call.'

The next day became the last working day for Pradeep Guha. I said, 'Pradeep, you are CEO of the company and I am not only chairman of the board but also the promoter shareholder. The minority shareholders will eventually ask me, and not you, about this disaster (if it happens). You and I have complete disagreement on this. Only one point of view can prevail. Hence you have to leave.'

Asking Pradeep to leave was actually not an overnight decision. It had been building up. His view on Zee's strategy to face competition was the breaking point between Pradeep and me. He could never earn my confidence. I had heard that he used to favour his friends within and outside the company. The film industry felt that the Zee Cine awards were biased, that they were given and denied to people who Pradeep liked or disliked. In one case his apparent favouritism had led to serious losses for our group. He made a deal with his friends at a music company on terms that in my estimate caused Zee to lose close to Rs 100 crores. When he asked me to increase spending without any assurance on success, I thought it was time for him to go. His departure was a blessing in disguise. My son Punit, who used to work under Pradeep, was promoted as CEO in 2007-08 and later also made managing director. Punit is now running the business prudently.

Two years later, my stand on not spending extra money was

vindicated. New channels like 9X folded up and others like Colors ran up huge losses. The third one, NDTV Imagine, launched with foreign investment but had to be sold—it was bought by Time Warner and Turner—because of its accumulated losses. Not just that, its new owners decided to shut down the channel. My twenty years of experience in TV has taught me a basic rule. Money does not buy eyeballs. The simplest of programmes can get ratings but the most expensive one can flop. Only those ideas liked by viewers do well, not those liked by programming heads.

THE POPULARITY OF some of our programmes, including the news bulletins, *Hamarey PMji* and *Aap Ki Adalat,* rose five to six months before the 1999 general elections. Rajat Sharma had invited Atal Bihari Vajpayee on *Aap Ki Adalat*, but he was refusing. I asked Vajpayee the reason for this. 'Rajat, *humse kuchh ultey seedhey sawal karega.'* Vajpayee was anxious about questions on his links with the Hinduja brothers and with his friend Mrs Kaul.

I told him that I would make sure he didn't ask these questions as these were about personal relations. The interview would focus on government and politics.

Vajpayee finally agreed and the interview turned out to be quite interesting. Rajat asked most questions, including Vajpayee's fondness for drinking. Vajpayee replied to all questions with an open heart. We made two episodes instead of one. When this was aired, I got a call from my friend Sharad Pawar, *'Aapne BJP ke 3 se 4 per cent vote badha diye hai.'* (You have increased BJP votes by 3-4 per cent.)

The BJP also took advantage of this interview and hundreds of thousands of videotapes of the show were distributed across the country. But Vajpayee and the BJP only grudgingly acknowledged the impact of the interview on the elections, which they won.

WE WERE POSITIVE about the growth in the media sector and had been talking to Tarun Tejpal's media company *Tehelka* for investing in it. The discussions were on but there was no decision on buying a stake. One day Tarun called me excitedly while I was in Singapore for an investors' roadshow. 'Subhashji, I have got something very big. The news is very big,' he said, unable to contain his excitement.

'What is it, tell me,' I said.

'No no, I can't tell you. You will leak the story. I can tell you only if you promise to air it on Zee News,' Tarun said.

I said we would air it if it was a correct story, not a plant, and if it was as important as he said.

He told me about the bribe taken in cash by BJP president Bangaru Laxman. In a sting operation, *Tehelka* had recorded Laxman taking cash. Jaya Jaitley, president of Samata Party, part of the BJP-led NDA government, had also been recorded similarly. I told Tarun that I must have the tapes at least two hours before the other channels. Tarun agreed. He sent the tapes to Zee News first and to all other news channels and newspapers after about 90 minutes. Our newsroom confirmed the sting.

I thought for a few minutes about airing this. Even though Vajpayee's inner circle was hostile to me, I did not want to hurt his government. In this sting operation two party presidents seemed to be taking a bribe. Government officials or ministers were not involved but it would have a big impact.

I was caught between my urge to show the sting while at the same time protecting the government from the repercussions. The news business is great for influence but also leads to situations where a tough balance has to be made.

I thought of a compromise. I decided to air it but it also pre-warn the Vajpayee government, I asked one of my executives, P.C. Lahiri, to inform Vajpayee's media advisor, Ashok Tandon, about the impending telecast of the sting.

I briefed Lahiri and told him that if Tandon asks us to stop the telecast, then we should say it was not possible. Even if we stopped, others would telecast it.

I also suggested that the best option for the BJP and Vajpayee was to suspend Laxman and Jaitley from their party positions until the *Tehelka* sting was probed.

Unfortunately, Ashok Tandon did not take this seriously and asked Lahiri to send the tape for his viewing. The contents were aired first on Zee News and half-an-hour later on other news channels. This sting operation exploded on the political landscape and created a storm for the NDA government.

Both Vajpayee and his deputy, L.K. Advani, were very angry with me and our group. Their finance minister, Yashwant Sinha, would later tell me that Vajpayee thought I wanted to bring down the government. Samajwadi Party MP Amar Singh had apparently fed this in Vajpayee's ear a few months earlier.

Even though all the media had picked up the sting and even the print media played it up, the BJP leaders held the airing against me. Sinha admitted to me later that he was asked to take whatever possible action he saw fit against me and my group. The government felt it could nail us on anticipated tax violations. But when Sinha asked his agencies to raid us, he got a surprise. Sinha's department of revenue told him that as per their intelligence, the Essel Group did not evade tax and did not have any illegal transactions. The revenue officials said that if a media house like Zee was raided, it would be futile and would embarrass the government further. But the officials thought of another way of hurting us. They told Sinha that the Congress government had launched a probe against us on foreign exchange violations. The officials told Sinha that they would not close the investigation and that they would intensify the probe.

I felt that Vajpayee's son-in-law, Ranjan Bhattacharya, and his principal secretary, Brajesh Mishra, did not like me. Both of them tried to turn Vajpayee against me. Our families had a relation for forty years, but Vajpayee became very unfriendly to me as result of their influence. They disliked me not just because my rivals were their friends but also because they believed that I was close to the top brass of the RSS.

20

FRIENDS AND ENEMIES

Taking on the Ambanis...reluctantly

I MOVED MY family to Mumbai around 1985. I did not have many close friends. There were only three people I could put in that category. These were Ravi Kiran Agarwal, Dev Ahuja and Ravi Ghai. Mukesh Patel was added to the list later. Ravi Ghai was from the family that made Kwality Ice Cream. His father was based in Delhi but Ravi ran the family-owned Natraj Hotel. It is Intercontinental Hotel now. Ravi purchased a paper-mill-cum-box-making unit. But that company was making losses. I bought the company from him, thinking it would fit in to Essel Packaging's plans. It did not turn out to be a wise decision.

Of them, my relationship with Mukesh Patel was the most interesting. It led to a war between between me and a prominent industrial group.

Mukesh Patel was a Mumbai-based entrepreneur. He was a car dealer and distributor. He was a dear friend though we did fall out for a while. Mukesh was involved in politics and spent a lot of resources and effort in supporting the Shiv Sena. He contributed about Rs 30 crores to the Shiv Sena for the Maharashtra state election. Soon, he became close to Balasaheb Thakre.

I advised Mukesh that joining politics or being too close to a political leader was not a good idea. You are spending money to buy a master, I told him. He would have to toe the line of his political bosses. This advice created a rift between him and me. He did not like what I said.

Soon Mukesh became so close to Thakre that some members of the Sena apparently started to get uncomfortable. There were reports that a person who had become too close to Balasaheb had disappeared in suspicious circumstances. After a while, Mukesh realized the danger he was in. He sheepishly came to me for advice. I told him to keep a low profile. However, as he wished, he was made an MP by Shiv Sena.

He was Gujarati and managed a community organization called Kelavani Mandal in Juhu. This mandal owned and managed several educational institutions. He was the managing trustee. More than 70,000 students were being served by the organization. Mukesh was friends with Praful Patel, who was then an upcoming politician.

Praful introduced Mukesh to Dhirubhai Ambani. Somehow Dhirubhai took a liking for him and they became friends. The Ambanis began to show some interest in the mandal that Mukesh was running. Mukesh told me that someone in the Ambani group wanted to take control of the mandal. This upset him.

One day Mukesh told me that he was going to fight out the Ambanis. I tried to dissuade him. He told me that Dhirubhai Ambani called him his third son. But now he felt they were trying to throw him out of an institution that was close to his heart. This mandal defined his standing in the community. I suggested to Mukesh that he should meet Dhirubhai directly and talk it out. I thought that Dhirubhai would stop whoever was trying to take over the mandal.

A few months passed before I got a call from Mukesh. I was in Kathmandu when Mukesh called me. He spoke excitedly and said, '*Bhai ji, hum Ambani ke petroleum tanker mein milawat ke liye maal lejate hue dekh liye hain.*' (I have seen adulteration being carried out in an Ambani petroleum tanker.)

He asked me to send a Zee News reporter with a camera crew to capture all this and telecast it. I told him to drop the idea as it would not lead to much. But he insisted. He was seething that the Ambanis had tried to grab his organization while pretending to be his friends.

For some time after that, I avoided his calls. I felt that the issue of adulteration had been around for a long time; there appeared to be no political will to stop it. And I had no intention of taking on the Ambanis for no rhyme or reason.

One day Mukesh almost raided my house at 6 a.m. and woke me up. He cried about what he was going through. Then he accused me of not being a true friend. He emotionally blackmailed me. I gave in and acceded to his request for recording the petrol adulteration for Zee News.

The Zee News team recorded the adulteration on camera. I asked Raju Santhanam, who was the editor, to check the story. And then I decided to hold the report and not air it. I was hoping that Mukesh would calm down and make up with the Ambanis. But Mukesh started pestering me to air the story. I kept avoiding him for another two weeks, hoping he would cool down.

I did not take his calls, I went out of town. I wanted to avoid a confrontation. But eventually, he got to me and I gave the go-ahead for the report to be aired.

The news report created a big buzz. The issue was raised in Parliament. Petroleum Minister Balram Jhakhar had a tough time. While the story died down after a while, it created a lifelong rivalry for me.

The Ambanis were upset with me for airing this news story. They apparently put me in their list of people to worry about.

As a result of my efforts to help Mukesh Patel, I seemed to have made enemies out of the Ambanis. I was not comfortable and was keen to end any misunderstanding over the report. The news report was not a personal attack on the Ambanis.

Murli Deora, an MP and a close confidant of Dhirubhai and his sons, wanted to put an end to this tussle with Mukesh Patel. Murli Deora told the Ambanis that if I was supporting Mukesh, then they must meet me. Deora told them that I did not care for too many people and was not easily intimidated. The only way out was to end the war between Mukesh Patel and the Ambanis.

So I invited Mukesh Patel and Mukesh Ambani to my house and requested them to patch up. I explained to Mukesh Ambani that I was only a victim of my friendship. If I am someone's friend I will support him without worrying about the consequences. I do realize that as my weakness, but that is me. The opposite of friendship is also true in my case. If people decide to fight me, without any provocation or fault on my part, then I do not back down either.

I asked Patel and Ambani to patch up and facilitated a dialogue between them. After some discussions they shook hands and promised to remain friends. The Ambanis agreed not to push him out of his Kelavani Mandal. All this happened in 1998-99.

Mukesh Patel had also started a business of importing foreign luxury cars under the tourism promotion scheme where duty did not have to be paid. Mukesh had either lost a lot of money in this business or he gave lots of political donations to various parties. As a result, he was in sizeable debt. One day he visited me with one Mr Rathi, who appeared to me to be a dubious character. Rathi tried to tell me how I could borrow money from banks and financial institutions without paying any interest. I was not keen on such schemes and showed him the door. I told Mukesh never to bring him to me again.

But Rathi did manage to convince Mukesh to invest in a gold-refining project. He told Mukesh that India imported 800 tons of gold every year, that they would import raw gold from mines overseas, and also collect lots of scrap in the Indian market, and refine both. This would be very profitable and would help Mukesh wipe out the debts.

A plant with a refining capacity of 200 tonnes was set up in Shirpur. The cost of the plant was about Rs 100 crores, but Rathi seemed to have persuaded Mukesh to overinvoice the capital expenditure. He inflated the fixed cost of the assets to Rs 350 crores. Then they raised project finance of Rs 350 crores from the banks. They both managed to rope in veteran movie actor Dilip Kumar to be the chairman. Mukesh asked me to find investors and I helped him find about Rs 50 crores as investment from my friendly investors overseas. After raising the money, Rathi seemed to have done some dodgy accounting to siphon off the money. Even Mukesh did the same to pay off some of his pressing debts.

But the company did not do well at all. Gold mines did not give the raw material to the refinery. The company did not have working capital. It became sick very soon. The banks put the receivables in NPA (Non-Performing Assets) classifications. During this time the doctors detected a rare ailment in Mukesh's nervous system. He died a few months later as a result of this ailment. He was just about forty years old. Some years later the banks sold the gold refinery to an asset reconstruction company.

The asset reconstruction company put the refinery on the block after a couple of years. We purchased the refinery by bidding for it and won it for less than 100 crores. I did this only to help the family of a close friend who was no more. I wanted to revive it, keep Mukesh's legacy and dream alive. It is still part of our group's operations.

21

LOSING $6 BILLION...

...as Zee's stock skyrockets, then crashes

WHEN THE MARKETS were rising in the late nineties, a share market trader, Ketan Parekh, started to accumulate the shares of Zee. He bought shares amounting to 8 per cent of the total equity. He started buying at Rs 100 and took it to almost Rs 300. There was much excitement about the stock in the market. In early 2000, the price shot up to Rs 1,500. Ketan Parekh started to meet me almost twice a week to discuss the business and check what we were doing. I could give only such information as was in public domain, but he could make out more by asking detailed questions. It had become almost a necessary evil for me to meet and discuss with him as he was the largest minority shareholder.

Mukesh Patel advised me to sell part of my shares in Zee as he thought that the price was artificial when it went up to Rs 1,500. Based on the share price of Zee, one day *Economic Times* wrote a front-page article stating 'Subhash Chandra is richest Indian.' I did not heed Mukesh's advice, though now I wish I had. Seeing my shares rise to these levels was a heady feeling.

I wanted to see the group become an institution. By then I had CEOs in place to run the day-to-day business. Even as I was enjoying the market success of the share, I got a shock when the price fell sharply to Rs 1,000. I became a bit worried and asked Ketan what was happening. I asked whether he was behind it.

He suggested that we buy back some shares together in partnership to boost the value. I did not want to get directly involved but told him that I could find friends who could buy. After four weeks of road shows in Indian and overseas markets we could generate demand. As a result, a number of brokers, institutional investors and banks bought the shares. Some of them bought in partnership with Ketan Parekh.

Ketan and other investors bought the stock at Rs 1,000, hoping that it would go to Rs 1,500 levels. Instead, the value fell to Rs 800. Now Ketan started putting demands on us. He said that if we did not support him, he would have to sell all his Zee stock. This would have further depressed the value of the share.

But the stock kept falling. Many people who were our competitors/detractors became active in pushing down the stock to accelerate the decline in its price. Some people who had been affected by reports on Zee News also joined in to batter the stock. As a result the stock was reduced to Rs 60. The fall between 1,500 to Rs 60 took less than twelve months. Within the period 1998-2000, the price had touched Rs 1,500 and then crashed to Rs 60 in 2001. The stock market itself crashed, particularly the new economy stocks, like dotcoms and media. Ketan Parekh could not support these stocks and almost declared bankruptcy.

We borrowed about Rs 150 crores to give Ketan to save him from becoming bankrupt. This would have affected us as well— if he had sold all his stock, the value of the share would have fallen further, reducing our market capitalization. We had to pay all this over a period of time to our lender. We have not been able to recover our monies from Ketan so far.

Though the stock market crash was a global phenomena, in

India it was blown out of proportion and termed as a stock market scam. The matter was debated vigorously in Parliament. The opposition was holding the ruling NDA government responsible. They demanded a joint parliamentary committee to look into two issues. One was the *Tehelka* exposé/episode and the other was the market crash.

The government took a political decision and agreed to appoint a JPC to probe the stock market scam, and not the *Tehelka* issue as that would have embarrassed the ruling BJP.

The JPC on the stock market scam became a free-for-all. Some lobbies worked hard to save influential stocks, while other lobbies tried to implicate their rivals. We were at the receiving end of some lobbies that tried to implicate us. The Zee Group's so-called involvement in the stock market scam became fodder for daily newspapers. Fortunately, no evidence was found about our role. The JPC exonerated our group and cleared our name.

The chairman of the capital market regulator SEBI, M. Damodaran, used to show much respect for me. He would say that I was a nationalist businessman. He would often call me to discuss issues. But he did not clear the regulator's case against us for more than four years, despite the favourable report by JPC.

SEBI sent us a show-cause on the volatile trade, accusing us of inside trading. I made several requests to Damodaran to end the SEBI case against us. I told him that the JPC had submitted its report and we were in the clear. But he kept the case pending. It seemed as if he did not want it resolved. I am yet to understand the reason for this.

Because of the delay in clearing the case, we could not raise money from the capital market for five years, from 2001 to 2005. We ended up facing a huge penalty for a crime that we had not committed.

When C.B. Bhave took over as SEBI chairman, I met him and told him about the matter. I requested that either we should be punished for our mistakes or the pending case be closed with appropriate comments. Keeping a case hanging did not make sense. He was reasonable. Bhave told me that the investigating officer was independent and was a quasi-judicial official. Therefore, it was beyond him to give any directions. But he agreed that it was reasonable to seek a final decision. Finally, the officer adjudicated the matter and we were let off the hook.

The entire stock market episode from Ketan Parekh to JPC to SEBI was a terrible experience for me. It took me ten years to get out of the mess. This not only caused huge financial losses for us as a group but it stalled our progress for ten years. Perhaps this is what 'the powers-that-be' wanted. We lost about $6 billion of market capitalization within months, as the Zee stock fell drastically. This affected our ability to raise funds and crushed the confidence of our team. Naturally, the business and our growth slowed down.

I learnt a big lesson from these episodes of 1999 and 2000: Never go near the stock markets, unless you are in that business as a trader. You should leave the share price of your company to the market and the investor community. You should ensure that you communicate your story and plans. You should answer the queries. But never buy or sell your own share because you think it is overvalued or undervalued.

But after that I decided to not to get intimidated by my detractors, including the Ambanis. I was upset that they had battered my stock by pushing it down, even though we had resolved the issue between us. Whenever any news came about the Ambanis, Zee News did not shy away from reporting on it. However, we had not gone out of our way to investigate the group even though we got many anonymous complaints against it.

In 2002, Amar Singh approached me on their behalf. He said, 'I want to ensure that you are friends with the Ambanis.' I agreed.

So I went with my brother Ashok and my two sons to meet Dhirubhai and both his sons. I told Dhirubhai, 'We are four brothers and between us we have nine sons. We are all together and united. I started with Rs 17 and even if I lose a lot, as I have, I will still be worth more than where I started, hence I am not afraid of losing wealth.'

Dhirubhai understood and made peace with me. I think he instructed his sons not to mess with me. I am in regular touch with Mukesh Ambani, more than with his younger brother Anil. I cannot claim that we are close friends but I feel I have cordial relations and that we trust each other.

22

IT'S NOT CRICKET

TV rights...and wrongs...and
a controlling board

THERE ARE SEVEN categories in broadcasting: news and current affairs, general entertainment, movies, sports, children's programming, infotainment and music.

We were in five of these seven categories.

We had decided to start a children's channel in 1997-98 but it was delayed because we lacked management bandwidth. In 2007, Turner Broadcasting, with whom we had a joint venture for distribution, suggested that we combine our four channels—two in English by Zee and two children's channels by Turner Broadcasting. This proposal did not work out despite efforts from both sides. Finally, our children's channel, ZeeQ, was launched in October 2012.

That brings us to the last category: sports. We did have a sports broadcasting business since 1998, but it was small and we wanted it to grow. The opportunity came in 2000, when the London-based International Cricket Council (ICC) invited bids for global telecast rights for the World Cup (and its other marquee tournaments such as Champions Trophy) for a period of seven years. The ICC has ten full members and a few associate members. It elects a president every three years. Jagmohan Dalmiya was its president at the time. We participated in the bidding process. To make a winning bid we had to hire a specialized sports law firm from London. We had to submit a letter of comfort and a bank guarantee of a hefty sum. We were technically qualified and when the financial bids of all bidders opened, ours was the highest. We had offered to pay $666 million and the next bidder had offered $550 million. Vijay Jindal, the CEO of Zee, was leading the process for us.

ICC President Dalmiya recommended to the managing committee that the rights be awarded to us as we were a qualified bidder and our bid was the highest. Meanwhile, the second-highest bidder, Harish Thawani from Bombay, approached me and said, 'Why do you want to pay such a high price to ICC? You and I can together take the rights at a much lower price.' I felt that it made sense to partner with the second-highest bidder; his bid was lower by over $100 million. Technically, under the bidding rules, it would have been possible to do so. I asked Vijay Jindal to meet Harish Thawani and see if the proposal made sense. But he brushed it aside and said, 'Sir, we will get the bid. Please do not worry about it.' So I left the project and the process to Vijay Jindal. Now, while writing this book, I realize that declining to partner with Harish Thawani not only threw our ambitions of sports broadcasting into a different orbit, but also created much disruption in cricket administration globally.

Thawani was a smart and shrewd businessman, yet seemed to like queering the pitch. When we opted away from a collaboration with him, I believe he went to Rupert Murdoch and warned him about the implications for him of Zee getting the rights. Apparently, there was a game plan to turn the tables on Zee in the process of bidding.

Meanwhile, Dalmiya's recommendation to award the bid to

Zee was not being accepted by the larger ICC board—the ICC is not bound to disclose reasons to the bidders. The board was asking Dalmiya to negotiate with all the bidders and to get a revised/new bid, which Dalmiya was resisting.

The board did not appreciate the resistance by Dalmiya. There were press reports at that time of misappropriation of funds by Dalmiya—the BCCI had become a rich board. Some of the members used these allegations to undermine Dalmiya's authority by raising issues of ethics and corruption against him. These accusations took Dalmiya and all cricket administrators by surprise and sent shock waves across the cricketing world. The whole cricket world was aware that before Dalmiya took over as president, the ICC's financial status was poor. He had been instrumental in reversing the finances of ICC and making it cash rich.

Three or four weeks later, we discovered that the bids were being renegotiated. I believe News Corporation played a role in getting Dalmiya removed as head of the marketing/managing committee that was overseeing the bidding. The ICC decided in its wisdom that Zee was not capable of fulfilling its commitments, though there were enough bank guarantees to back our bid.

Finally, the ICC awarded the rights to Global Cricket Corporation (GCC), a News Corp entity which had in turn engaged World Sport Nimbus, which was a joint venture between the World Sport Group of Singapore and Harish Thawani (Nimbus), to act as an agency for further syndicating the rights for the next seven years to various broadcasters. GCC, represented by World Sport Nimbus, had to sell the broadcast rights of these events to broadcasters in India and across the globe, for which they again approached ESPN-Star Sports (a joint venture between News Corp and ESPN), Sony and Zee in 2002, before the 2003 World Cup (Sandeep Goyal was the Zee CEO at that time). The India rights finally went to Sony.

IN 2004, THE BCCI invited bids for telecast rights of all their cricket events in India for four years. Here again, Dalmiya was in a decisive position as BCCI president. However, lots of surprises were in store for me.

The BCCI had to give the rights to a sports broadcaster. When the bids were opened, our bid was the highest. I called Dalmiya, whom I considered a friend, as a courtesy, and spoke to him in Marwari. I told him that Zee had bid for the rights. He was warm initially, but the moment I said our bid was the highest, he turned cold. 'I don't know about this, we will see,' he said and hung up on me. This took me by surprise.

I thought that Zee Sports would naturally be awarded the rights since we were the highest bidders. But after this call, I realized there was something wrong. It turned out that Dalmiya, probably using technicalities, wanted to award the rights to ESPN-Star Sports and not to us. Our bid was for $260 million for four years and we had added another $21 million, taking the bid to $281 million, subject to the BCCI allowing us to revamp domestic cricket (read: launch the Indian Cricket League). The ESPN-Star Sports bid was for $230 million for four years (which was the bid duration as per the tender document). ESPN-Star Sports had put a rider that if they were given the rights for five years, then they will be willing to pay $308 million to the BCCI.

After talking to practically everyone who was influential in the cricket establishment, I came to believe that Dalmiya ran the BCCI as his private company. He seemed to forget that its main function was to promote cricket in India. He took the word 'Control' in the name of BCCI (Board of Control for Cricket in India) too literally.

The BCCI called both the highest bidders—Zee and ESPN-Star Sports—for a meeting at Chennai in September 2004, about three weeks after the bids were opened. My colleagues and I appeared for a discussion at the offices of India Cement, the company owned by N. Srinivasan, an important member of the managing committee and a confidant of Dalmiya, and a man who would go on to become BCCI president and ICC chairman himself.

As the meeting started, Dalmiya spoke to us in a hostile tone. I maintained my cool but realized that we were fighting a tough battle. By evening, we got to know through our intelligence that Dalmiya seemed set on giving the rights to ESPN.

I realized I would need help from other people for taking on Dalmiya, who I felt was being unfair. I began to call up the people I knew. I happened to have very good relations with BJP leader Arun Jaitley. I have a good equation with him even today. Jaitley was the president of Delhi & District Cricket Association and an important member of BCCI management. I sought Jaitley's help but he was cold and said, 'I will see, Subhashji.' Much later, I learnt that it was not Dalmiya but Jaitley who favoured ESPN-Star Sports because he genuinely believed that the foreign companies are more honourable or professional than Indian ones. If Jaitley was in the government, as he is now, would his beliefs remain the same? Sometimes, these genuine beliefs don't fade away.

I also knew Rajiv Shukla, who was representing the UP cricket association. I tried to speak to Rajiv, but he did not take my calls. I lost my cool and gave his wife Anuradha Prasad a piece of my mind. I told her that they had forgotten about the time when they needed my help and when Rajiv Shukla was working for me. I warned her that Rajiv's behaviour would not benefit him in the long run. Rajiv was reportedly supporting Dalmiya. Now I don't know how true this is, but some people allege that Rajiv is open to persuasion and can change his position based on practicality. He was an unknown journalist who rode to fame on the back of the Zee network. He had hosted a show called *Ru-Baroo* that helped him connect with political leaders.

While Jaitley and Rajiv Shukla let me down, I had unexpected support from inside the BCCI. N. Srinivasan was sympathetic to our plight. Few people knew about it. My friend Ravi Parthasarthy, MD of Infrastructure Leasing and Financial Services (IL&FS), who knew Srinivasan, had asked him to try and ensure that Zee got a fair deal.

As it developed, Srinivasan's efforts were not enough to deter Dalmiya from backing ESPN-Star. During the bidding process, all of us were staying at the same hotel in Chennai. I was not only frustrated but was also very angry. I decided to take the matter in my own hands. I called Dalmiya in his room at night and spoke to him in strong language. Short of threatening him, I conveyed my anger and rage at what seemed to be his partisan behaviour. Dalmiya was rattled. He did not expect me to call him directly and be so aggressive.

Dalmiya realized that I would not give up without a nasty fight. Next morning, he complained about me to the managing committee and told the entire board about my call and my accusations. This served my immediate objective, since my gameplan was to publicize Dalmiya's bias.

When the matter was heard by the managing committee, some members realized the gravity of the situation and supported us. The committee wanted a fair decision. It asked Srinivasan to step in to mediate between Dalmiya and me. As part of the settlement, Srinivasan asked us to increase our bid further. My colleagues and I were then called in front of the committee and were told that ESPN-Star Sports had agreed to drop the condition of five years and now were willing to pay $308 million for four

years instead of five years. However, in an attempt to seem fair—since we were the highest bidders initially—Dalmiya had to call us and give us the opportunity to match and increase our bid to $308 million for four years. Dalmiya thought that I wouldn't accept this but I immediately called his bluff and agreed to increase our bid amount to $308 million for four years. This rattled him and he then put forward another demand—that we would need to pay Rs 100 crores within the next 48 hours. My colleagues rightly argued that this wasn't a condition in the bid document but I stepped in and immediately agreed to that unreasonable demand as well. By now, Dalmiya understood that I would not let go of this contract at any cost, at which point the meeting ended with the telecast rights being awarded to us.

Dalmiya took this loss of face badly. He did not respond when I called to apologize and make up with him. I called him an hour after the rights were awarded to us. He coldly said, 'I cannot see you as I am already on my way to the airport,' and hung up. Actually, Dalmiya seemed to want to continue to fight against me, though we had ensured that Rs 100 crores were paid within 48 hours. Possibly encouraged by Dalmiya, ESPN went to court, challenging the award of rights to Zee. Finally, the rights awarded to us were cancelled by the Bombay High Court.

But I don't give up easily, especially when I believe in something. We gave Dalmiya a hard time in the courts. The Supreme Court passed many strictures on the BCCI based on our petition. But Dalmiya knew that courts would not go beyond passing strictures.

I took on Dalmiya in another way. When the next election for BCCI presidentship came up, I went out of my way to support Sharad Pawar. He was elected as BCCI president while Dalmiya lost.

Everyone who was against Dalmiya worked with me to oust

him. But helping Sharad Pawar win did not help me at all. He turned a blind eye to the fact that we had worked hard for his win. Rather, he helped our competition. Truly, there are no permanent allies or enemies in the world of the BCCI.

I GOT A chance to further my interest in sports when, in 2006, I was approached by an investment banker. He told me that that Sheikh Abdul Rahman Bukhatir, a UAE businessman, wanted to sell half his stake in Ten Sports. I had known the Sheikh for long and decided to buy that stake. Unfortunately, we did not negotiate well and I paid much more ($50 million for our 50 per cent stake) than that business deserved. Since the acquisition of Ten Sports, we have kept bidding for cricket telecast rights and won in Pakistan, Sri Lanka, South Africa, West Indies and Zimbabwe.

Meanwhile, in the year 2006-07, the Indian cricket team was performing very badly. Its ranking was just above Bangladesh's. On analysing the reason, it became clear to me that India was not performing consistently because they did not have bench strength. The entry of new players was close to impossible in the Indian team. To make cricket more interesting and revive the sport, we decided to launch league cricket. I was keen on partnering with BCCI for this since I thought it would be good for it to promote a new form of cricket.

The 20-20 format idea had been around with me since the time I had bid for World Cup rights. In 2004, when I had bid for Indian cricket rights, I had included a complete chapter in my bid with details of starting a domestic league. We had offered to pay extra money ($21 million) to the BCCI for just allowing us to launch it. But the BCCI never even looked at the concept, and Dalmiya had just dismissed it.

When Pawar became president of BCCI, I thought of reviving

the concept, and explained it to him. But he remained non-committal. 'See if it works,' he told me.

Later, I learnt that he had misread what I was trying to do. He thought that instead of helping the BCCI, I was trying to compete with it by launching my own league. He not only did not help me, but directed other members/office-bearers of the BCCI to do everything possible to stop me from succeeding. The reality was I was not trying to create a parallel system to the BCCI. I wanted a format that would give a platform to youngsters, add some excitement to the game and, in the process, create a talented bunch of new players for the national team managed by the BCCI.

We went ahead and launched the league even without the BCCI's help, and created several teams. But the BCCI, feeling sidelined and insecure, advised (even threatened) players, advertisers, and sponsors against joining our league. Their member associations refused to give us the cricket stadia for the games.

Deprived of official cricket grounds, I approached state governments and asked them for help. We got a stadium in Gurgaon and another near Chandigarh, both owned by the Haryana government, and one in Hyderabad, owned by Andhra Pradesh government. Later, we got a stadium in Ahmedabad.

We managed the first 20-20 league event, in December 2007, very well. The audience response was great. The rating and sponsor reaction were good. But we could recover only 5 per cent of the event cost of Rs 100 crores.

At the next ICL event, again in 2007, we recovered about 30 per cent of the cost. In 2008, when we were planning the third event, there was a global economic slowdown and crisis in financial/capital markets. All the advertising money in the market dried up. So far, it was my personal money being invested in the league. We had invested about Rs 300 crores of our own money in the first two events as we did not want our listed companies to be affected by the failure of the league. But by 2008 we had run out of personal free cash flow. I could borrow but decided against it because of market conditions. After that, we did not revive the Indian Cricket League. Meanwhile, the BCCI started the Indian Premier League and used its might to make it successful.

The ICL suffered because the establishment did not support it. But more importantly, it was hit by the market crash. If the downturn had not happened, even the IPL would not have been able to prevent the ICL from being a success. We would have broken even and made profits in another two years. It was the economic downturn which killed the ICL, and not the BCCI as is popularly perceived. Had this crisis not hit, we would have recovered about 70 per cent of our costs in the third season and 100 per cent in the fourth.

The sports broadcasting business is over Rs 4,000 crores a year. The time is still good for this. We had filed a case against the ICC for not recognizing the ICL. Under ICC rules a private organization could launch its own tournament and apply to the ICC for unofficial cricket status. The ICC then supplies the umpires and other facilities at a cost. But the ICC refused to accept our request. Kerry Packer had also challenged the ICC when he started one-day cricket. The ICC later patched up with him and Cricket Australia gave him telecast rights for a long period of time on his terms.

My fight with the ICC and the BCCI is still on in London and Indian courts. We have a case pending against the BCCI in the Delhi High Court. We have also challenged the actions of the ICC and the ECB (the England and Wales Cricket Board) in the courts of London. We also have a case against the BCCI in the CCI (Competition Commission of India).

In retrospect, I made a mistake while launching ICL. I did

not sell the league teams. I owned all of them. I was overconfident. If I had sold them, I would have raised instant capital. At that time I was keen on making ICL a success before selling the teams. I actually wanted to list each team as a separate company.

Today there is so much illegal betting in cricket. If the teams had been listed, all the betting would have become legal via the stock exchange. If the team was doing well, the stocks would rise, and vice versa. This system would have created a new format and a healthy environment for cricket.

23

CALMING INFLUENCES

Vipassana brings peace...but turmoil, too

I WAS A curious person from childhood. I had lots of questions about human existence, about God and so on. I came in contact with hundreds of saints and gurus from various sects and beliefs. I practised the path suggested by many of them, but I never found satisfaction nor did I get answers to my questions.

I found a totally new way of thinking and living when I met Acharya Satya Narain Goenkaji, who became my guru. He was the elder brother of a family friend, Gauri Shankar Goenka. I had heard much about him from family and friends and finally met Acharya Goenka at his residence in Juhu in Mumbai in 1989.

Acharya Goenka introduced me to Vipassana, a form of meditation. He told me about his own past and how he discovered Vipassana. 'I grew up in Myanmar and was a successful businessman. I was also involved in social activities of the Indian community and used to head several non-profit organizations. I used to lecture on the Bhagwad Gita. While speaking I would get deeply involved in the emotion of the stories of Lord Krishna and Lord Rama. So much so that I would start crying and my audience would also weep with me.'

Goenkaji suffered from severe migraine and all attempts to

cure it had failed. He had received treatment for it across the world, in places like Europe and Japan. Nothing had helped. He used to take morphine injections to dull the pain.

This went on until he was introduced to and taught Vipassana by Sayagyi U Ba Khin, the first accountant general of Myanmar and also an authority on Vipassana. Vipassana was a technique taught by Gautam Buddha to his disciples. It was preserved in its original form by disciples and passed from generation to generation for more than twenty centuries.

S.N. Goenkaji said that he was initially hesitant to adopt Vipassana. According to him, the Hindu Brahminical view of Gautam Buddha and his teachings was negative. (I discovered, though, that many Hindu scriptures give Buddha a god-like status.) Since he had been brought up in a religious Hindu family, he was sceptical about the benefits of Vipassana. Despite his hesitation, Goenkaji undertook a Vipassana course for ten days. He realized that it did not teach anything that went against Hinduism.

The great outcome of the course was that Vipassana not only helped him get rid of his migraine but he also found his true calling. He started practising it regularly under the supervision of his teacher, Sayagyi U Ba Khin. Goenkaji proudly displayed the picture of his guru at his home.

Sayagyi told Goenkaji that the technique was discovered by the Buddha in India. 'This is a gift from India to the world and I feel indebted to it. I wanted to repay my debt to India, but I cannot go there. Hence you go and spread Vipassana there. This will lift the debt India has on me.' Sayagyi passed away in 1971.

Following his guru's advice, Goenkaji left Myanmar for India in the early 1970s to teach Vipassana. From just about ten to fifteen students in the 1970s, there are today more than 5 lakh practitioners. There are hundreds of Vipassana centres in India and abroad.

After listening to Goenkaji with rapt attention, I approached
Vipassana with an open mind. He suggested that I try it for
about ten days. The first three days were very difficult for me. In
the first three days, the entire concentration is on breathing.
You just observe your breathing: inhaling and exhaling. The
focus is on the nostrils. The mind wanders so it's hard to
concentrate at first. All types of thoughts and memories flood
your brain. But after a while, the mind starts to calm down and
one is able to concentrate for a longer period on the breathing.
This is called Anaapana.

I managed to complete the first course of ten days. I felt not
only light but could experience sensations taking place inside
the body from head to toe. At one point of time I felt as if my
body did not exist; it felt like air. This feeling and experience
was just for a few seconds or even less.

At the start of the course, the staff at the centre plays recorded
audio and videotapes of Acharya Goenkaji in which he says,
'You have come here for ten days. It is like being operated upon
in a hospital, hence follow the instructions of living the life of a
monk—no sex, no make-up, no theft, no lies and so on. Also,
you will strictly not speak to anyone other than your teacher or
the Dhamma worker (volunteer) for any assistance. Don't even
make eye contact with others. If you violate any of this or break
the course in between without completing it, you can
dangerously harm yourself.'

When I came back after the course, something had changed
in me. People in the office felt that I was a different person and
they saw a glow on my face. I felt good.

I wanted to return for another course, but got busy with
work. I could not attend for another eight months. But after
that I have become fairly regular. Now I sit for one long session
at least once a year and try to manage smaller three-day courses
during the year.

Vipassana made me calm and relaxed. I could reflect on several issues that were deeply ingrained in me without my realizing them. I was able to control my anger. I started to understand the real cause behind what was happening to me— whether it was happiness, sorrow, anxiety, excitement, desires of wealth, sex and so on. It seemed to me that I was progressing towards finding the meaning of life. All the knots inside me were being unlocked. There were times I felt that my body did not exist. I tried to touch it and feel it but it was not there during those moments.

Over time my confidence and faith in Vipassana has increased. It has relieved me of a lot of stress I face at work. I try to practise it everyday but am able to do so four days in a week, at home or office. This helps me improve my temperament and allows me to be more patient.

My elder son Punit took two or three courses and my wife Sushila tried it a few times but did not take to it. I recommended it to my father also. He went for a course once and enjoyed it. But on his second try he felt that Goenkaji was fuelling anti-Hindu sentiments in his recorded discourses. After that my father stopped. My younger son Amit, my daughter Pooja and my daughter in-law Shreyasi went for the ten-day course but could not last more than two days and returned. I recommended it to my three brothers but only Ashok took my recommendation and managed to take the ten-day course. I was keen that my mother also take Vipassana but somehow it did not happen.

Goenkaji told us that he does not call himself a Buddhist. He said Buddha never taught Buddhism. In all literature, there is talk only of Dhamma. Nothing else. He never gave any name to Dhamma. For Buddha, Dhamma was your nature and behaviour. To follow your Dhamma is to be a good Hindu, a good Muslim, a good Christian and so on. Buddha never believed in any 'ism'.

I continued my practice and kept meeting Goenkaji. When

Zee TV started in October 1992, he requested that a few of his discourses be aired on TV. I was glad to do it. As it happened, the TV discourses helped the spread of Vipassana. Goenkaji also acknowledged it. I would get calls from him for some sundry help or the other and I would attend to it. I would attend Goenkaji's discourses and participate in the activities.

Then a big project related to Vipassana came my way. Sometime in 1995-96, Goenkaji called me and said he was looking for land around Mumbai. He said, 'I want to build a replica of the famous Shwedagon pagoda of Myanmar in India. The original Shwedagon is a solid structure. But I want to make a dome-shaped hall that looks like the Shwedagon pagoda. I want to build a place where 10,000 people can sit at a time and do Vipassana. I am keen to repay the debt of Myanmar and my teacher Sayagyi.'

He said, 'Gautam Buddha had said during his lifetime that his remains (Dhatu) should be enshrined in small quantity inside as many pagodas can be built. These should be built at a crossroads inside or near the habitation of large populations. His Dhatu will create strong vibrations in and around the pagodas which will motivate people to adopt Dhamma.' Several relics of Buddha are housed in temples across the world.

A public trust, Global Vipassana Foundation, was formed and I was nominated its chairman. I offered to help Goenkaji. Not only did we donate a large piece of land within Essel World, but I also dedicated myself to the task of building it with *tan, man aur dhan* (body, mind and wealth). The pagoda construction, started in 1998-99, took twelve years of continuous work. Milllions of tons of stone, lime and red mud were brought from Rajasthan to Mumbai. We decided to build a stone structure as per Goenkaji's wishes—so that it would last for many decades. Cement or steel was not used for building the pagoda; only a pure stone structure can outlast normal cement. This unique structure has a dome which is six to eight times bigger than the biggest existing dome in the world. I was lucky to have been instrumental in building it.

The pagoda was inaugurated by the president of India on 8 February 2009 in the presence of the who's who of India and Maharashtra. I was the master of ceremonies at the event. Goenkaji was happy like a child who had found his lost mother.

Sadly, the joy did not last too long. A terrible bout of infighting broke out between some of the followers of Goenkaji. I was the chairman of the Global Vipassana Foundation, the trust that owned the pagoda. Some people decided to denigrate me and push me out. I got a rude shock when a situation was created for me to resign as chairman. About a week later, I received a further blow. I was accused of trying to control and own the pagoda as my personal property. Some people went about saying, 'Subhash wants to throw everyone out.'

It is my belief that the conspiracy to malign and push me out was hatched by Goenkaji's sons. They convinced my friend Ratnakar Gaikwad and a few others to oppose me. I do not have any proof, but my hunch is that Vallabh Bhansali, a leading investment banker in Mumbai, was also made a party to this without his realizing it.

I got such a shock from this episode in 2010 that I almost went into depression. I made the further mistake of not sharing this with anyone in the family, including my wife. I used to cry like a child at night, especially when I was away from home and family. Vipassana was like oxygen to me. My association with Goenkaji and the construction of the pagoda had given new meaning to my life. Now it was being taken away by a few petty minds.

I not only stopped associating with the pagoda and with Goenkaji, but I also became lethargic and would not take much interest in business or family affairs. My brothers and family thought I was sad because of the separation amongst us brothers, which also took place around the same time.

But the real cause of my depression was that I had been separated from a person who had taught me the life-changing technique of Vipassana. He had been my support for over two decades and had prevented me from making serious mistakes. How could he do this to me? Goenkaji knew that I had not cheated anyone. Why should anyone think that my intention was to usurp the pagoda? I had donated land for the good cause of Dhamma. Not for personal control.

Such was the impact of this episode that I faced physical and mental illness. I suffered an attack of pulmonary embolism while I was in Cannes, France, for a small vacation with my friends. Pulmonary embolism is caused by a clotting of blood; it can be fatal if it occurs in the heart. I was taken in an air ambulance to a nearby military hospital for treatment. I remained hospitalized for three days.

I suffered a lot during this period of eight to ten months. I swallowed the pain and the humiliation. I bore the loss of respect for Goenkaji silently, without letting anyone else know. Somehow Goenkaji understood what had happened. Perhaps the power of Vipassana made him realize his blunder. This type of character assassination was against the teaching of Buddha.

After a few months, Goenkaji wrote me letters inviting me back to the foundation and to be with him again. He also sent me messages through common family friends. We belong to the same family tree of Goenkas. I met him and made peace with him. He resumed calling me for sundry help and tasks. He also asked for my help in obtaining Buddha's relics from the possession of Government of India.

At the beginning of 2012, Vallabh Bhansali came to my office and said, 'Subhashji, a big wrong is likely to happen in the Vipassana trust. Guruji wants to make his son Prakash Goenka his successor.' My response was cold.

'It is our Dhamma duty to save the Vipassana organization from becoming a family trust. You have been largely responsible in spreading Vipassana in India and also you have benefited from it,' he said. 'You cannot shirk your responsibility.'

I remained noncommittal and said I would think about it and revert. But I spent a few sleepless nights thinking about the problem.

I had overcome my depressive behaviour with the help of Vipassana meditation. I had been motivated by Dhananjay Chauhan, secretary to Goenkaji, and Arun Toshniwal, honorary secretary of Global Vipassana Foundation.

Moreover, I had a responsibility towards not leaving my family in debt. I had borrowed during the separation between us brothers. I had pulled myself out of the depression so that I could work and repay all debts to the family balance sheet. In the business separation I kept most of the corporate liabilities to minimize the burden on the businesses that my brothers would handle. This pressure to reduce the corporate liabilities kept me going.

It took me some weeks before I could come to a conclusion about whether I should interfere in the succession plan at the foundation. Opposing the plan would mean going against the wishes of Goenkaji. Finally, I decided that I had to step in and at least give my unsolicited opinion to my teacher's plan to put his son, who had not been practising Vipassana, in charge of the trust.

My reasoning was simple. Goenkaji had stopped me from going down the path of *adhvogat* (sinning). Now Guruji was in trouble. His son was manipulating him, taking him down the wrong path. I wanted to save him the same way he had saved me.

Converting a public trust, meant to help millions, into a family-run foundation was immoral. The trust was to be run by Vipassana practioners. Ideally, the best among the followers should have taken over the foundation. A dynastic rule in the foundation was abhorrent. Guruji treated everyone equally while his son had created a feudal atmosphere. His son saw himself as a leader with many followers, just because he was a biological son of Goenkaji. He had no other merit. This was against all concepts propogated under Vipassana.

I had no choice but to convey my views to Goenkaji. I took the support of some Dhamma brothers, including Ratnakar Gaikwad. Even though he had helped throw me out of the foundation, Gaikwad was an honest and dedicated Dhamma worker. I also enlisted Bhansali's help. We voiced our disagreement to Goenkaji, saying, 'You cannot turn this public trust into a family trust.'

Goenkaji resisted all those who supported our argument. They included students, teachers, Dhamma workers and trustees of the foundation, including some members of his extended family.

At one point, besides Gaikwad, he wanted to make Priyanka Gandhi and a daughter of former president Pratibha Patil as the office-bearers after him. But I was told that Priyanka distanced herself from the foundation after learning about the changes.

I write all this with a heavy heart and in the hope that the power of Vipassana will bring order to the whole issue. Goenkaji passed away in 2014 at the age of eighty. Just about six months before his death, he apologized to me during one of his discourses. I felt terrible and small about the apology. A teacher shouldn't have to apologize to a student. I wrote him a letter offering my own apology for opposing him. Sadly, my words didn't reach him. I was later told that this letter never reached him, though it was delivered at his residence. Hopefully, the foundation will be back in the control of genuine Dhamma workers. I continue to practise Vipassana regularly, though I did not attend the course for some time after these incidents.

Vipassana remains very important to me. It calms me down. It helps me introspect and manage tough situations. Most importantly, it keeps me equanimous. This is the key for mental peace. In success I am not too exuberant while in failure I don't get so upset.

BY 1997, WE were distributing our network of channels/ programmes globally in about fifty countries, which grew to about 160 countries by 2011. The question was: what next? Once, when I was working on future programming for Zee channels, I had a thought about promoting our ancient Indian system of health and well-being. I wondered why no one was showcasing our ancient medical knowledge.

I commissioned some shows to promote traditional medicines. We invested about Rs 30 crores on these shows in 1997-98. We created programming under the brand name 'Chakra'. The shows were innovative and inventive. In one show we analysed cosmetics to assess their positive and negative effects and showed the viewers that they contained many chemicals. Some of the chemicals were harmful to the human skin, but were included to give the cosmetic in question a longer shelf life.

In another show, we demonstrated how to make cosmetics from kitchen supplies and plants from the kitchen garden. There were shows on ayurveda, its use and history.

But not much came out of it. The shows did not do well in India or abroad. I wondered what we could do next after catering to the Indian diaspora market. We realized that Zee had to consider doing niche programming for global audiences. We knew that we could not compete with the Western media and entertainment companies in general entertainment programming.

In early 2000, we thought of going in for wellness-oriented programming, not just for the Western market but globally. Till then, we were catering to the South Asian diaspora, but had not ventured beyond that. I thought that wellness programming would give us access to the global community.

I studied the US market and started work towards launching a wellness network, and launched it under the brand name 'Veria' in 2009. Veria is derived from the Latin word Veritas (the Truth). This was the only channel of its kind launched for an American audience, made by a company with parentage of Indian origin. We decided that the anchors, the language and the content should be totally suited for American viewers, and hence the creative work was entrusted to Americans.

The Americans have developed a lot of faith and interest in ayurveda, nature cure, yoga, pilates, etc. They treat it like alternative medicine. We were keen to cash in on this need, especially because most of the knowledge comes from India and eastern countries. Unlike our channels in India, Veria was privately held and not listed on any stock exchange. Moreover, none of the listed companies of our group was involved.

Launching a channel in the US was tougher than we realized. Despite our launch in 2009, we still don't have adequate distribution. We were seen in only about 15 million out of 110 million TV homes in 2012. Now, in 2015, we have reached over 30 million homes in the US. The cable operators and DTH companies say they don't have space or bandwidth for our channel. They seek carriage fee in millions of dollars. Strangely, they don't charge this fee from the strong networks of American companies. The US needs solutions in health care but there is hardly any corporate social responsibility among the distributors of TV networks. They hide behind shortage of bandwidth but are happy to air adult (pornography) channels.

We have not given up. The households that are exposed to our programming are happy. We will succeed, hopefully soon. The channel was renamed as Z Living in 2014.

I HAVE THREE children, two sons and a daughter. I feel I have not been a great father, at best only a good one. I have not spent enough time with my children to become their close friend. Or to be as close as a father and his children should be. I have a good relationship with them but I would not call it great. I do feel sad about it at times. At the same time, I am happy and proud of all three of them. They are humble and helpful to everyone. Sometimes, when they help the undeserving, I get angry. But they don't throw their weight around.

It's difficult to compare my relationship with my children with what I had with my Dadaji. I would tell him anything and everything. But it's not the same between my children and me. I also do not share (as is my nature) much with them and perhaps that is the reason of their being reserved.

There are two reasons for this, though not excuses. One was the mistake of sending them to a hostel, at a very young age. I later realized that it was a mistake to keep them away from us. The second big reason, of course, was the time I had to spend at work and business.

I sent my two sons to the hostel so that they could learn the hardships of life. I wanted them to learn to cope on their own, away from the comfort of their home and pampering by us, but that was wrong thinking on my part.

When they returned from Switzerland, I realized I had lost them somewhere. They were in Le Rose school in Lausanne. My daughter spent time at home but due to my work, I could not be her close friend.

When Punit was in Switzerland, he was distracted from his studies. This was not only felt by me but I also picked this up from conversations with the kids around him. Perhaps even Amit mentioned something to this effect.

When they were kids I would make an effort to spend time with them—we would go out of town for a week to ten days, whenever I got time. We would take a holiday together twice a year. But as they grew older, these holidays reduced. Now they have their own lives and families and do things their own way, which is right. We still try to be together once a year.

They share more with their mother, but still not much. They are closer to Ashok than they are to me.

When they came of age, I wanted to train them personally for the business. Both the boys worked with me for a while on different projects.

My thought process was simple. They should observe me at work as I had observed and learnt from Dadaji when I was young. But I am not sure whether they were keen on the same method.

In 1994, Amit wanted to set up a data centre. Those days information technology had become a booming sector and everyone wanted to enter it. I allowed Amit to go ahead. I also gave him a simple piece of advice. If too many players are entering the sector, then it will become a commodity. And once that happens, the margins fall and it's tough to make profits. However, Amit was quite convinced about it, and got the money he needed as initial investment.

Amit also started working in the education company. For a while, he complained about the CEO. But I did not react. One day he sacked the CEO and informed me. I backed his decision. I did not want to interfere. It was important that he decided on his own and took responsibility for it.

Punit was placed in the satellite project Agrani. He reported to the CEO, Jai Singh. Jai was a tough boss and he groomed Punit for almost three years. Punit did his graduation by correspondence course. He preferred to learn from experience and not in the classroom. He was not keen on academics, so I thought it was better that he started working.

Dadaji often used to tell me that people were making a fool of me and making money from me. Sometimes, I think my sons too are gullible and that some people take advantage of them.

Punit is an outgoing personality and likeable. He makes

friends easily with a diverse bunch. But he is a bit carefree and not very disciplined. The success which he has achieved in the entertainment content business as CEO of Zee Entertainment has given him confidence and helped him improve as an entrepreneur. He brought stability to the company. The flagship channel has come from the fourth position to being number 2. Amit, on the other hand, is not so outgoing and he has his set of close friends. He is a clear-headed thinker and is driving the businesses of gold refinery and infrastructure towards success. I had many friends when I was young. But those friendships were with the people I dealt with in businesses at different times, and I kept those relationships. We helped each other. I am not sure about the friendships Punit and Amit have developed, nor I have tried to find out.

I kept them protected as much as I could from the influences of growing up in an affluent family. I would put an expenditure limit on Amit when he was doing his masters in the US. They had all the basic amenities. They also had credit cards, but they had to watch their expenditure.

Of all the three children I was closest to my daughter Pooja. She was the youngest and as is natural, closer to her father than the mother.

She wanted to study in London since a couple of her friends were going there, too. I was not keen to let her go. But she insisted and I had to permit her. Born in 1979, she is two years younger than Amit and about four years younger than Punit.

She is very strong-willed and proud. When she did not do well in her graduate studies in London—she had taken home science and had to reappear for some exams—she came to me, cried and apologized. She said she would pay for the costs of study in London. I wanted to laugh but I cried, too. Then she started to study in Mumbai.

She married a boy of her own choice. Now they live in New

York where my son-in-law is in the real estate business. Initially, I was against the marriage. The boy was not of any means, but that did not affect me.

I met his mother and also discussed the future with him a couple of times. I thought he was not right for my daughter. For one he was about ten years older than her, and I did not find any clarity in his thoughts as to what he wanted to do in life. In my discussions, I found that he would criticize all and sundry but had a high opinion of himself. This was not a healthy approach to life, I thought. But Pooja was keen and claimed to know him very well. I gave in to her wishes. She is happy now. And when she is happy, so am I and my wife Sushila.

Punit as well as Amit had arranged marriages. Initially, Amit wanted to marry a girl of his choice. She was a Brahmin. I told Amit that he should marry someone who could accept and like our family culture. I met the girl and told her about my concerns. I told her that her approach to life and thinking was different from ours. Therefore, she may not be able to fit in. I explained our family's customs and traditions. She understood and realized that she would have to compromise her way of living. We are a fairly conservative and traditional family. Both Dadaji and my father believe that a Brahmin girl should not marry into a Vaishya family.

I explained all this to Amit. They both agreed to call off their plans to marry.

There are certain values we still hold dear in our family. You will not find a Brahmin working as a domestic help in our house. He can be a chef, but not a server. I would never wittingly hand over my used glass or bowl to a Brahmin server even in someone else's home.

24

THE ART OF MANAGEMENT

Staying focused on the present

PEOPLE OFTEN ASK me how I got this ability to make friends from an early age. My personality in my youth was very different from what I am now. I was chatty, confident and usually had a smile on my face. I would be open about my problems and feelings. There was a sense of directness and honesty in my interactions. And I would not hide my ambition to do better. I didn't give an impression of being a lazy person. My readiness to work hard and my confidence were infectious. Many people who helped me in my teenage days did so only because of my confidence and earnestness. I was ever willing to help others. I think this helped since I managed to earn the faith of people who helped. Some people promise to help but don't follow up. I tried to be there for whoever needed me.

There was also a mismatch between my age and maturity during my younger days. My expectations and ambition were so obvious that it endeared me to many people. I think people tend to admire someone who exudes energy, positivity and an urge to do better. This is not a common attitude. Many young people these days are easily satisfied or are cynical. If they face a problem, they blame the circumstances but do not work to change them.

I was always restless. Everytime my business stabilized, I wanted to try something new. Perhaps this is an inherent personal trait that I can't explain. For me trying something new is exciting. I become dogged and totally consumed by a new project. I seek thrill in the new ideas that I explore; the joy of a challenge, the excitement of conquest over a challenge.

I don't like my money to sleep; it should not lie idle. I want it to be working for me or be used for a good cause. I don't like to hoard money.

I prefer to enter segments that others ignore, especially if these appear impossible to break into. My entry into rice exports, broadcasting, packaging, amusement parks are all examples of this. I don't mind failing if I know that I tried my best. The Indian Cricket League and Agrani were my pet projects. I invested money, time, effort, resources into them. They did not work out, but I have no regrets.

I still work more than those younger than me. The word impossible is still not there in my dictionary. Once I am convinced, I develop a strong strength of conviction, and the courage to follow through on my commitment.

My English was never very good but I was not ashamed of speaking in broken English. Over time, I corrected my mistakes and improved my language. If I had been concerned about my wrong English, I would not have been able to speak good English at all.

I am not scared of jumping into situations that would scare others. I do not fear anything. This was the legacy of my Dadaji. He would say that death comes once. Why be scared of something that comes only once.

Being restless is a natural trait, but this can also be an acquired attribute. We should continue to strive for something better. We should keep trying to create a more challenging goal for ourselves. That's how I drive myself all the time.

When I see someone not performing, I am frank enough to tell the person that it's not working out. I request him or her to leave or change jobs within the group. But I see many of our senior colleagues, including my brothers, sons and nephews, empathetic towards non-performers. They don't want to face the issue. They tend to become comfortable with such people and they get protection. They tend to choose people who become personally loyal to them rather than to the company. I think it's important to be professional about such matters. Protecting a non-performer is not good for the business and also the person being protected. This is unprofessional too. The non-performer may be in the wrong job and thus not doing what he or she is best at doing. Empathy that results in protection would lead to a negative result for the employee as well. He or she might be better off in another job within the group or elsewhere.

Managing so many companies is not easy. It requires a methodical approach. A few years ago I started a system and process of meeting all the important executives in the group. Once a year I interview and discuss with them their issues with the company, or with their boss, and also try to suss out if they are the right fit for the job they are currently doing.

I meet the CEOs regularly. Then I meet all the department heads that report to the CEO. And then I meet the people who report to the department heads. All this happens at periodic intervals.

I go through their HR feedback forms. I also read the feedback forms of the people who report to these senior executives. This way I get to keep a track of what is happening in each company and also how each executive is performing. We call this whole process Samvad or dialogue.

One has to keep one's eyes and ears open. And the feet should be on the ground. For example, at Zee Learn, I share the feedback I receive from parents and employees with the company's CEO and HR head. In fact, I tell all CEOs in our group to be more engaged with their teams. I once received a complaint from a former employee that his accounts had not been settled. That was not the right thing even if there were issues because of which the person had been asked to leave.

In my mind the real art of living as well as management is to be focused on the present. My motto is: *'Bhootkal vyakul kare, ya bhavishya bharmaye, vartaman mein jo jiye so jeena aa jaye.'* (The past makes you restless, the future deludes, but living in the present is true living.)

In the last two decades, I have changed a lot and so has my management style. I used to be very short-tempered but this has reduced. With anger come other problems. You say the wrong words and take the wrong decisions. You also end up misjudging people. You become irrational. I am still not out of it fully.

I have changed my thinking in another way. When I would bring a senior executive on board, I expected him or her to earn my trust. I realized that this approach does not work. Now when someone joins, I begin by trusting him or her. Only when things go wrong does my trust erode. I learnt all this from various management books I read over the years, and also with my own experience with people. Vipassana also helped me. I used to delegate responsibility once a person earned my trust. Now I delegate work right in the beginning.

Take the example of my colleague C.S. Vishwanathan. He has handled my work and other corporate work in Delhi for many years. But matters were not being resolved in a timely and accurate manner. Everytime I assigned him some work, he would not complete it in time. And he would have one reason or another for it. He seemed to be negative about every task. One day, in exasperation, I told him to leave. He was angry with me as well; he left my room and started to pack up.

After about 20 minutes I calmed down, and so did he. I asked him where he would go at his age of forty-five-plus. He did not have an answer but said that he had no choice because of my behaviour. He said I was constantly scolding him.

'I get angry because you start every conversation by saying that the task is not possible. If I call you on Sunday, you tell me it's a holiday,' I said explaining my behaviour. 'I don't ask you for work unless I have to. Let's try one thing. For the next three months, I want you to approach things positively and make a sincere attempt to do everything I ask you to. Even if it seems impossible, try it. Tell yourself you can do it.'

Vishwanathan tried it. And it worked wonders. Now he finds solutions to the most difficult issues and I have no reason to be angry with him. The lesson I learnt was that instead of getting angry, one must have faith and guide a person who is struggling.

One reason I have been able to succeed is that I have had truth by my side. It has caused some problems too. I had to fight with some political leaders. I was known as a quarrelsome person in some circles. Many people avoided me as I was known for using sharp words. But I have always said what I felt, even if it angered people.

The book that impacted me the most is *First Break All the Rules* by Marcus Buckingham and Curt Coffman. I read it about ten years ago. I also had long conversations with management gurus like Ram Charan and the late C.K. Prahlad. I shared some of my business ideas with them. My son, Punit received mentoring by Ram Charan a few years ago. We also started to get advice from him for the group's growth and strategy.

To me there is no difference between traditional and modern management philosophies. I can summarize it simply. Select the right person, define the right outcome and empower them. These words evolved over the years. It was like this from the beginning but I never articulated it.

Senior executives must have the freedom to be entrepreneurial. Nobody has been sacked for making mistakes. But I have been quick to ask people to leave for being unethical or non-performers.

Each organization has a culture, ethos, belief. The management style may change but the ethos does not.

For instance, the most important ethos is to not break a commitment, whether there is a written agreement or not.

I keep a diary of my observations about the CEOs and the people who report to them. It takes up a lot of my time to keep meeting the business heads and interacting with them. But as the group chairman this is my job. I don't have to manage the day-to-day activity. I enjoy going through the revenue and other numbers once every month. At the same time, I have my ear to the ground, so I know the micro-picture in order to manage the macro.

Sometimes people have tried to create a new culture that did not sit well with ours. Then I had to intervene. When Pradeep Guha joined us, he brought a lot of people from his previous company. A new coterie was created. That itself was not bad. CEOs should have the freedom to hire. But he believed his team was superior to the existing old executives. His team could do no wrong and others were always at fault. I don't agree with this attitude. A leader has to take everyone along. A leader has to extract the best out of every employee. Condemning people on a selective basis is not the solution.

Managing within a family framework is also important. After the initial separation of work within the family, we did some further operational division of work in the year 2000. This was to ensure that our children knew what businesses they would manage.

When my brothers and I met to discuss the division, the entire meeting would have lasted not more than an hour, and there were two more meetings of 30 minutes each. We worked out and wrote down what everyone would get. Some of it had been demarcated earlier. But we agreed fairly quickly and smoothly. We did not involve our sons or any outsider in the meeting. This was a division of responsibility but not separation. I still feel as if we are together and the businesses work as one group.

We decided that each of us would hold a small amount of shares in the other brother's company. We have decided that we will not sell these shares to anyone except the brother who owns the company as per the arrangement. Thus if I own shares in Ashok's company, I will not sell these without asking him first.

This division was also about succession planning. Many business families do not do it. If it is not done in time, it gets ugly. Doing this has helped our group grow. My sons and brothers are not anxious about their role and future. They take total responsibility for their respective business and its development. The younger generation was getting anxious about its future. They were keen to know what part of the business they would handle. It was important to carve out the responsibilities.

I realize that in the early days I would take many decisions without consulting my brothers. Sometimes they would come to know about the decisions from others and not from me. This was not totally right. In some ways, it was my responsibility to drive the growth in those days. But I think I should have been more inclusive.

Some other business families were impressed with the division in our family. I helped some families in their business division and succession planning. It was a bit like my Dadaji helping other families during their times of crises.

Now I have some direct responsibilities and some indirect

ones. Jawahar, for instance, is helping me manage even those businesses in which he may not have any shareholding, in addition to Dish TV and others where he has a stake. There are a few small businesses that my sons and Jawahar's sons are handling together.

THERE IS A practice in the Agarwal community to donate 10 per cent of their profit to charity. I have tried to observe this since I was a child in Adampur and Hisar. The practice is called *Dasaundh*.

This used to be mentioned even in the accounts ledger of the traders. This money would be spent on sadhus, *gaushalas*, or any other charitable work such as getting rooms constructed in ashrams. My father built *gaushalas* for abandoned cattle in Adampur and Hisar. There are several thousand cows there.

The belief is that the traders will earn ten times more money than what they have donated during the year. The charity money will be given through the year and debited to the charity account. The profits will be known only by the end of the year and 10 per cent of the profit for the year will be credited to that charity account. We donate at least 10 per cent even when we are not sure what the profits will be. Our family continued to donate small sums even when we were in debt. Once it is known that a businessperson donates for worthy causes, there is no dearth of choices. Many voluntary organizations seek you out for donations.

When I moved to Delhi, I donated money for the Maharaja Agrasen Hospital being built in Punjabi Bagh. When our Dadaji died, we started a girls' school and college in his name in Hisar. My father believes we should not publicize the charity done by us, hence you will not find our names mentioned in this regard in any insititution. The pagoda we built cost more than Rs 100 crores in land and donations but we did not add our name to it anywhere.

In 1993, during the Christmas and New Year vacations, a group of people from our Agarwal community from Mumbai, including my family, went deep inside the rural and tribal areas of Bihar, in what began as a Vanyatra. We were visiting the habitats of the Adivasi community. These people subsist on basics but are happy. The trip was hosted by Shyamji Gupt, a social worker and reformer.

I sat with my family in the hut of an Adivasi family. We could see that all they had was the clothes they were wearing. Tied under the roof were various dry leaves and dried vegetables and sundry odd items. We were told that during the rainy season they survive on these dried food items as they don't get food supplies for three months.

I felt terrible observing them. Compared to them our lifestyle was criminal. We buy clothes even though our wardrobes are full. Our kids do not eat if they do not get their favourite dish. But this family was surviving on the bare minimum. Still, somehow this Adivasi family looked happier than all of us.

There was a gut-wrenching conversation with them. I saw a cow outside the hut with her calf. *'Bhaiya barsat mein aapke bachhon ko gai ka doodh to mil* hi jata hai na,' I asked. I assumed that their children would get cow's milk during the monsoons.

The answer stunned me and I felt very small. *'Sahib aapki maa ka doodh aap aur aapke bhai behno ke alava kisine piya kya,'* he asked, adding, *'To hamare bachhe gai ka doodh kaise piyenge, woh to gai ka bachhda he piyega.'* I was touched by their values. The Adivasi family did not use their cow's milk as they felt that its rightful beneficiary was the calf, not the children of the cow's owner.

This conversation humbled me, and I hope my family also remembers this. I decided to do something for them. I began to contribute to Ekal Vidyalaya, an organization run by Shyamji Gupt. The concept is to create single-teacher schools embedded in the community. One teacher is appointed per school and that teacher teaches children from nursery to class 5. There are no school campuses or structures. The classes are held under a tree or at the house of the teacher or, in some cases, the gram panchayat allows the use of a room/hall made for the community. By 2014, more than 50,000 such schools were run by Ekal. There are close to 5,000 volunteers who either work free since they have other sources of income or do so for a small stipend (called Maan Dhan).

A young adult from the local community is trained to be a teacher. The cost per school per year was about Rs 8,000 when it started in 1988. Now it is about Rs 20,000. I have been part of the Ekal Vidyalaya for many years as a donor and a board member. Now this foundation has also expanded its activity into health care. It helps people become self-reliant. We have dreams to take it to the next level as part of developing rural India. This is a difficult task, yet doable.

This example of Ekal is one of many such wonderful and social endeavours. India has millions of people who are engaged in non-profit organizations in various fields. Giving is a more enjoyable experience than taking. People who do not know how to give or only know 'how to take' are miserable much of the time.

In business, most people are in a transactional mode. For everything you give, you expect something in return. But in philanthropy, there should not be any expectation of any benefit. There is just a deep sense of satisfaction and happiness.

Hughes did build the satellite but at specifications much lower than what we had demanded. They sold the satellite to another company in the Gulf, called Etisalat.

Sometimes I wonder why some US companies are so unethical. They do what they feel like and their government supports them. They want to manipulate their clients also.

After Iridium failed, the consortium that owned it was ready to sell us the project for just $10 million. But I was not interested in it. Instead, I thought of launching our own satellite for TV. We even lobbied with the government to allow a private Indian company to own a satellite. This policy was announced in 1996-97. We are perhaps the only Indian company registered to own a satellite company.

FIVE SATELLITE MOBILE phone projects were launched in the early 1990s. The first one was Iridium, led by Motorola, with the participation of many developing countries, including India. The second was by International Communication Organization (ICO), led by a consortium of telecom companies from different countries, including VSNL of India. The third was Agrani (US), the fourth was Global-star, and the fifth, Etisalat.

ICO had launched a few medium earth orbit (MEO) satellites but had gone into bankruptcy. The shareholders had put it on the block through a US auction process as ICO's registered headquarters was Delaware. I was keen on satellite communications so I decided to bid for it and went to Delaware. I was so obsessed with buying a satellite company that I was ready to try anything. When I tried to bid for it, the ICO management did not support us. The VSNL representative initially did not support us, so I had to get someone in the government to call him.

We had long meetings with the management but they did

not think we had the financial clout or wherewithal to turn around the company. Our bid was supported by the Enron Corporation. Its India CEO, Sanjay Bhatnagar, had flown in with me. The global CEO of Enron, Kenneth Lay, was supporting us. Those days Enron was seen as a big and influential US company.

The other bidder for ICO was the US entrepreneur Craig McCaw. He had created and launched the cellular phone technology in the US and sold it to AT&T. When McCaw saw that Enron was backing us, he decided to make a deal with us. We decided to buy the satellite telephony company in partnership. We agreed to give him $40 million and invest more and the deal was done. But before I could make the entire payment, the Indian stock markets crashed. This created a serious problem for our finances. I met Craig and told him that it would be tough for me to pay him. I requested him to return the money I had invested in the deal. But he returned only about $20 million, half of what we had invested. The rest we got in ICO shares that we sold later for a much lower value of $4 million.

The satellite industry is a very small close-knit group. We had almost become a part of the group. However, by this time cellular phones had become popular so I told my CEO, Jai Singh, that there was no point continuing with the satellite dream. Jai Singh had worked with us for four years and Punit had worked under him.

I was keen on mobile phones, but by the time the satellite story ended, it was too late. I decided to keep my focus on the media industry.

25

A SATELLITE PROJECT RUNS AGROUND

Losing big time in a high stakes game

WHEN WE STARTED Zee, we did not know what satellite meant. The chief engineer of the state-run Doordarshan gave me a simple and effective description of a satellite. He said it's a bit like a big mirror in the sky. Whenever a signal or a beam of light is sent up, it reflects it back within a certain area. That's the footprint or reach of that satellite. When it is geo-stationary, it means that it is orbiting the earth at the same speed as the rotation of earth. This means its position in the sky is fixed in relation to a point on earth.

When Rupert Murdoch bought a stake in our company and became our partner in 1994, we wanted to gainfully deploy the money. I think it's in my destiny to spend money every time I get some cash. Many believe it is not healthy—at times even I feel so!

This is what happened in 1994 as well. A senior executive in Star TV, Michael Johnson, approached me and suggested that we build a satellite together. He know I had made money by selling stake in Asia Today Ltd to NewsCorp.

This satellite would offer mobile telephony with a small

hand-held phone that would connect directly with a satellite. Around that time, the Indian government was initiating the process of handing out mobile phone licences based on the GSM technology. This required a network of towers across the region. I was excited by Johnson's proposal as I thought that a single satellite could be more efficient than a network of hundreds of thousands of towers. That's how the mobile phone operators work currently.

Johnson suggested that we discuss this with Hughes Communications. The idea was to get Hughes to build a satellite that would offer telephony in low telecom density areas in Asia and Africa.

We launched the project and called it Agrani (staying ahead). The company was named Afro-Asian Satellite Communications. We set up this company in London. Apart from Michael Johnson, Steve Moss from Star also joined and they started putting together a techno-economic feasibility study, which needed to be credible enough to get financing of $600-700 million.

While the report was under preparation, we operated from my residential apartment at Baker Street in London. I stayed in London for the project report and also made several trips, mainly to the US, to get basic details for the satellite. I had to wade through all kinds of technical and commercial data.

To own and manage a satellite required several steps. We sought permission from the International Telecom Union (ITU) for an orbital slot for our satellite. We also sought permission from the Government of India for private ownership of a satellite by a domestic company. However, India did not have such a policy. Thus we had to look for other countries where we could do so.

At first we had thought of setting up the company in Mauritius. But the minister of telecommunication wanted a larger fee for the government than was viable for us. We had planned an investment of close to $1 billion. But despite the scale of investment and request by the Mauritian prime minister himself, the telecom minister did not approve our proposal. The project would have put Mauritius on the telecom industry map.

Disappointed with Mauritius, we went to Gibraltar. A friend in London, Joe Bellotie, was from Gibraltar and offered his help. The government of Gibraltar supported the project and we filed for the orbital slot with ITU from there. But we faced obstacles from the UK government. The ITU demanded consent from the UK as Gibraltar was a territory of the UK. This proved to be tough for us. As a result we had to return to our own country.

The techno-economic feasibility report was done at a cost of close to $15 million. I paid a hefty salary to both Michael Johnson and Steve Moss, apart from half-a-dozen senior and junior management staff. There were huge travel costs and other expenses for a period of almost twenty months. Once the report was ready, we decided to work out the details for its implementation and financial closure. But before I could take a step forward, I got a shock. Michael Johnson disappeared with the only copy of the report.

He reportedly went and sold this report to AT&T in the US at an estimated price of $30 million. I had not been aware of this but Johnson allegedly had a history of duping people.

Unfortunately, I had not done enough paperwork to claim total ownership of the report, even though I had funded it entirely. Johnson did not stop there. Not only did he vanish with the report, he tried to monopolize the project's rights. We had applied for a licence to run a satellite phone project in India. But Johnson filed cases against us, claiming intellectual property rights over the project, calling it his personal knowhow. He claimed in an Indian court that the idea was his. This case, which went up to Supreme Court with Kapil Sibal appearing for us, was finally won by us.

I proceeded with my plan to set up the project. We broadly knew what was to be done. Steve Moss continued to work with us and we hired an excellent scientist-cum-technical director, named David Greenwood. We also hired a full management team headed by Ted McFarlen, a former Hughes executive. I got the Essar Group to invest in the project. I also brought in some domestic financial institutions to take stakes in the project.

I invited bids for building the satellite with the specifications prepared and written by Greenwood. The document of commercial terms was prepared by Steve Moss. Only two companies—both American—were capable of building the satellite at the cost we wanted. One was Hughes and other was Lockheed Martin.

The specifications and detailed terms and conditions of the satellites were discussed and agreed with both bidders at a meeting in New Delhi's Hotel Surya. Both bidders were requested by Agrani to give their offer for building and launching the satellite on a turnkey basis, including the ground segment. I had made a three-member committee for the bidding process. This included Harish Aggarwal, who was a friend and an electronics engineer from BITS, Pilani. Our chief engineer Greenwood was responsible for technical evaluation. I camped in the hotel with my team while both the companies sent their delegations for the process, which had been decided in advance.

Since Ted McFarlen had worked for Hughes in the past, I had not involved Ted in the final price negotiations for the $700-million bid. I did not want his judgment to be affected. We would have raised the amount required. Of this we would pay $550 million for satellite manufacture. The rest was other related costs. I had a suspicion that Ted was still very close to the Hughes Communcations. I did not fully trust him. I used to wonder whether he was planted in my team by Hughes. But I had no proof of it.

One evening after a detailed evaluation of the financial bids received, we called it a night at about 1 a.m. in the hotel. By that time I had pretty much decided that Hughes's bid was higher than that of Lockheed Martin. I had indicated to Hughes that the contract was likely to go to Lockheed.

Then, around 3 a.m., Ted called and woke up my colleagues, including Harish. Ted told Harish that I had slept after drinking too much and was feeling unwell. Ted added that I had authorized him to attend the final negotiations and proceed with the bid process. Ted then asked my executive, 'So what is the final bid of Lockheed? I want to know so that I can deal with them tomorrow.'

Somehow, my executive could not believe that I had gone to bed so drunk that I could not handle further negotiations. He came rushing to my room at 4 a.m. to check on me. He told me what Ted had said. I could figure out right away what was going on.

I called Ted immediately. 'You are sacked from this moment. You are no longer the CEO.'

Ted did not have much to say to me as he was guilty, but he alerted the entire team of Hughes about this.

By 5.30 a.m., my old friend Vijay Dhar was in my room. I was a bit surprised. Ted had called the India CEO of Hughes group, Pradyuman Kaul. He was a relative of Vijay Dhar. Kaul had asked Dhar to persuade me to change my decision on the contract.

A little later, Kaul also came into my room. It was still early morning. Kaul asked me about my decision.

I told Dhar: 'Sir, I can't give the order to Hughes. They have played a dirty game with me. They tried to find out the rival's bid by using the CEO with whom they had old connections. They are cheats. And their price is higher, too.'

Hughes had sent Dhar to negotiate on their behalf. Dhar

said: 'Hughes wants to apologize for this. They will reduce their price. They will charge $20 million less than what Lockheed is asking.'

The fact was that while the Lockheed price was slightly better, its technical specifications were lower. The Lockheed satellite would not have served our needs. If Lockheed were to match the Hughes specifications, the cost would have been about $70 million higher.

Actually, I had played a smart game with Hughes. That night before going to bed, I had told Hughes that the bid would go to Lockheed, just to check their reaction. I had not said anything to Lockheed.

This precipitated the sequence of events that exposed Ted and Hughes's game plan.

But with Dhar and Kaul, I continued the charade. I told Dhar, 'Bhai saheb, Please don't put pressure on me. But let me talk to them.'

The Hughes team came at 8 a.m. with a fresh offer. They had reduced the price and added a few more features to the technical specifications. This deal was even better than original and far better than Lockheed's bid.

That afternoon on 21 May 1995 we signed the deal with Hughes for $67 million.

Soon after, the construction of the satellite began. We began by paying about $10 million to Hughes. After I fired Ted McFarlen, I hired an Indian CEO with a good pedigree, Ashutosh Garg. But even he was influenced by Hughes. Within four months, he started speaking the same language as Hughes. He tried to convince me to dilute the specifications of the satellite. He argued that we should give up some of the features of the satellite so that it does not delay the delivery. But our chief engineer Greenwood would not agree with it.

If we diluted the power of the satellite signal, then it would

26

STABBED IN THE BACK

Steeling myself as the system turns
against me

IS THERE NO respect for pioneers in our country? I write this chapter of my life with a sense of pain and anguish. I have been targeted in the past for various reasons. But the attack on me by steel magnate Naveen Jindal caught me by surprise.

In 2012, Jindal made an elaborate effort to entrap senior editors of Zee News in a bid to retaliate against me. Naveen wanted us to stop reporting on the national coal scandal that was dominating headlines in all the media. Opposition parties were badgering the UPA government on the issue of companies benefitting from an opaque, preferential system of coal allocation by the Congress-led UPA government. They were quoting the figures put forward by the Comptroller and Auditor General (CAG). Naveen's Jindal Steel and Power (JSPL) was the frontrunner among the companies that seemed to have benefited.

Naveen is the scion of the O.P. Jindal family and their multibillion steel business empire. When we were growing up, Mr Jindal was our inspiration. From a humble beginning, he created a sprawling empire and a successful political career.

It all began on 17 August 2012 when the CAG tabled a report

in Parliament on coal blocks allocation. The report said that the process of allocation had been non-transparent. If the blocks had been auctioned in an open competitive manner, the government would have earned lakhs of crores. In any event, as it stood, the loss was estimated at Rs 1.78 lakh crores.

Like all other media, Zee News began reporting on the issue from the day the report was tabled. CNN-IBN, PTI, ANI, Times Now, *Hindustan Times*, *Indian Express* and many others were constantly reporting on the scam. By the end of August, Zee began a series on all the companies indicted by the CAG report and the coal scam. As the matter developed, deeper linkages between companies that received coal blocks and their political connections were revealed.

In September that year, the BJP declared war on the Congress-led UPA government on the coal scam issue. All newspapers, magazines, channels were devoting significant amounts of time and effort in reporting all aspects of the issue. The political, business, policy and regional dimensions were being covered.

Out of fifty-seven coal blocks allocated to private companies, seven had been allocated to Naveen and his family. But the proportion of coal in them was higher. Apart from Naveen, coal blocks had been allocated to his brother Sajjan and brother-in-law Sandeep Jajodia. CAG figures showed that Naveen and his family had been allocated about 50 per cent of the coal calculated according to the reserves. So it was natural that there was extra focus on JSPL and Naveen. CNN-IBN reported on 6 September that Naveen had been allotted more blocks than anyone else.

Naveen was not just an industrialist. He was also an MP representing the ruling Congress party. Channels played up this fact and pointed to possibilities of preferential treatment to Naveen.

I was in the US when the coal scam was being reported. One day I got a call from Naveen's uncle, Sitaramji. He said there was a negative report on Naveen on Zee News and requested me to have it dropped. I told him that I would check and get back to him. Then I called the Zee newsroom to check the status and found out about the entire coal scam coverage. I called Sitaramji and explained the situation. And I told him that it was nothing personal but part of the scam coverage.

Sitaramji understood and I thought the matter was over. But it appears that Naveen was very angry with Zee News for its coverage of the scam and the focus on JSPL. Zee pursued the coal scam longer than others. Perhaps that's what irritated him. Or perhaps he saw a motive when there was none.

On 10 September, Naveen was at a function at Le Meridien Hotel in New Delhi when a Zee News reporter caught up with him. The reporter asked Naveen for some clarification on the coal issue. Instead of answering, Naveen became angry and pushed the cameraman away. 'Zee News is showing all rubbish on its channel,' he said. After manhandling and abusing the Zee News team, he stormed away.

For the rest of the day, all channels telecast his misbehaviour. The next morning, the newspapers reported it as well.

After this incident, the Zee News team redoubled its efforts to investigate the activities of Naveen's group companies. I heard about this incident in the US and supported the decision to increase coverage of Naveen if the stories had merit.

Naveen realized he was in trouble. This public attack on the media was turning people against him. He swung into damage control mode almost the same day. He thought the first step was to buy peace with Zee News.

The corporate communication head of JSPL, Ravi Muthreja, called the head of Zee Business, Samir Ahluwalia, and met him at a coffee shop the same evening. Muthreja asked Ahluwalia to stop airing all reports on Naveen's involvement in the coal scam. He offered a bribe to Ahluwalia for stopping the stories.

Ahluwalia was angry and refused, saying that it was impossible since the reporting was being done in public interest. Muthreja then made another attempt to stop the stories by making an offer to the channel. He said that Naveen was willing to spend Rs 25 crores on advertising on Zee's news channels; it would be a quid pro quo for stopping the stories. Ahluwalia thought that Jindal trying to stop stories by offering an advertising deal would be a story in itself. He told Muthreja that he would discuss it internally and get back to him.

The next day Ahluwalia met Zee News editor Sudhir Chaudhary, and told him about the incident. Both decided that the only way to nail the attempt to bribe Zee would be to get the contract with Naveen's company. With a signed copy of the contract, they would use it as proof to further indict Naveen. To my mind this was not a good idea to begin with, as our staff were not equipped to do a sting operation of this kind.

In the ensuing discussion between the Zee editors and the JSPL team, the latter asked that for at least four years, no negative news be carried against JSPL. In return, JSPL would offer Rs 25 crores to Zee every year, or a total of Rs 100 crores. The two editors had a few more meetings with representatives of Naveen to get the contract signed.

During the course of the meetings between 13 and 19 September, the editors met Muthreja and other representatives of Naveen. The CFO of Naveen's company, Sushil Maroo, asked Ahluwalia again to take a bribe and settle the issue. He refused. Meanwhile, Zee continued to report on Naveen and even sent a team to Chhattisgarh and Odisha, where a number of mines mentioned in the CAG report were located. For the stories, the news team kept asking Naveen's team for clarifications and carried their response. It was normal, responsible behaviour by the news team.

Naveen must have become worried about this. Between 22

and 26 September, Naveen called me several times and sent text messages on my mobile. He wanted to meet me on the issue. I told him that I would meet him on my return from the US by the end of October.

Naveen kept telling me that Zee News was showing fake information about him. 'Whatever I am, it is because of you. Please stop the stories,' Naveen pleaded. Some of it was true. Our group had helped Naveen a lot in the past. Naveen was close to my brother Laxmi. When Laxmi was in charge of Zee News, the channel would have done over 200 reports in Naveen's favour. Laxmi would indulge him also because we had old family connections with the Jindal family.

Despite our support to him in the past, Naveen had turned against Zee News. While he was contacting me in US, I had been told by my team about his attempt to bribe Ahluwalia.

On the other hand, Naveen told me that my team was asking for a bribe. I told Naveen emphatically that my team would never ask for a bribe. But he kept pleading with me to stop the stories. I became irritated and asked him to stop bothering me. I told him to discuss the matter with my brother Jawahar, who was in Delhi.

Meanwhile, I was getting messages and alerts from friends warning me about Naveen. I was told that he was capable of anything to get his way. There were reports about Naveen's attempt to silence Right to Information activists who were asking uncomfortable questions.

Jawahar called me and told me that Naveen had been in touch. Naveen even asked his mother to call Jawahar and invited him to their place for a meeting. Jawahar met Naveen, his brother Ratan and their mother. They pleaded with him to stop the stories on the group, invoking the many decades of association between the families. Jawahar agreed and called me to convey the message. I was not in favour of it, but I agreed to stop the stories since Jawahar was asking me.

I called the newsroom with a heavy heart and asked them to suspend the reporting on Naveen for a while. I thought this would put a stop to the Naveen episode. Little did I know what Naveen had been planning.

A few days after Jawahar had called me and I had stopped the reports, an FIR was registered on the night of 2 October 2012 at 11 p.m. (despite it being a holiday) in Delhi against the Zee News team and it included my name. Jindal and his company accused Zee of extortion in the FIR. Newspaper reports started appearing, claiming that Naveen had caught Zee editors while they tried to take a bribe. It turned out that when the editors had been meeting and discussing the advertising contract with Muthreja and other executives of Naveen, the latter had been secretly recording the conversation. The FIR lodged was based on the doctored version of the sting.

Now Naveen's devious plan became clear. He trapped the editors into the advertising contract offer while he got Jawahar to stop reports on the coal scam. For an outsider, it was easy to believe that Zee had stopped the stories because of the promise of a Rs 100-crore advertising deal.

The Delhi Police seemed extra eager to investigate the matter. They sent notices to the editors and me, asking us to appear before them.

Naveen held a press conference on 25 October claiming that the Zee editors had asked for money in return for stopping the stories. Naveen played edited extracts of the conversations to give an impression that the editors had asked for money and advertising contract in return for stopping stories.

The nightmare began after that. It appeared that the entire government, the Delhi Police and the media were against me and the Zee group. The editors were arrested by the Delhi Police. I was called to the police station and questioned.

I met S.B.K. Singh, joint commissioner of police of crime

branch, Delhi. He was aggressive with me and asked if I was capable of murder. I met Home Secretary R.K. Singh and asked him why the police were so aggressive towards me. Singh said he had heard the recording and was convinced that Zee was guilty. I told Singh that he had heard only 14 minutes of selected conversation that was longer than 6 hours. But my logic was not appreciated by the police officers or the home ministry. The stand taken by Zee News on coal scam and Naveen remains vindicated. The case against Naveen continues as the CBI is still probing him and has not dropped any charges against him.

I realized soon enough that this was an elaborate attempt to destroy my reputation and business. I don't know who is responsible for this but I was told that some people in the Congress were angry with me and wanted to teach me a lesson.

A few years earlier, I had hired Aditya Sinha as the editor of DNA newspaper. I was cautioned by people from the Congress that Aditya was a bad choice as he had written against Sonia Gandhi and I was told that DNA would become a platform for his views. The Congress did not want Aditya to use DNA to write against Sonia Gandhi. I told people from the party that I would not let him be unreasonable. I also said that if he went somewhere else, no one would be able to control him.

I am not sure why the Congress supported Naveen on this issue. Did Naveen donate a large sum running into thousands of crores to Congress? I don't have the answers.

I was equally disappointed with the media houses. All of them turned against me. Instead of seeing the case against Zee as an attack on media, other publications treated this as an exception.

The fight between Naveen and me was not a personal fight. It was a fight between a corrupt system and a media company.

I find this difficult to reconcile with. Is it because my media group is much younger than others? Are other media owners jealous of my success? *The Hindu, Hindustan Times* and *The Times of India* group started more than a hundred years ago, while Zee is just over two decades old.

I helped the media industry by launching a private broadcasting network against all odds. I helped all private media barons by creating the path to a vibrant broadcast business. Then why this animosity against me?

As I have said publicly, I have reason to believe that this FIR was registered in the capital of India against a respected media house because it had the blessings of Sonia Gandhi and Rahul Gandhi. There were others in Congress party such as Digvijay Singh, general secretary, Congress party, and Manish Tiwari, spokesperson and minister of information and broadcasting, who supported this action. When a friend of mine spoke to Prime Minister Manmohan Singh, the latter regretted his inability to help, though he was sympathetic towards us.

This was an unjust act by the UPA. In response, I personally supported Narendra Modi's campaign for prime minsitership.

I would like to think that I pioneered some change and contributed to the media industry. In any other country, I would be treated as an icon or a hero. Somehow, our country does not want to recognize or respect its pioneers.

The attack on Zee News will be a dark episode of my life. My pain will remain.

27

NEW HORIZONS

Paving the road to success

AFTER THE STOCK market upheaval in the year 2000, our media business lost almost 90 per cent of its market value. It took about five years for stability to return to it. The businesses of the Zee Group had been broken into four different companies and they were doing reasonably well. But I realized that the group needed some diversification to spread the risk for the future. Apart from packaging and real estate, most of the group was in the services sector. As a strategy of growth, I felt the group should enter a stable, long-term business rooted in the fundamentals of the economy. Unlike my previous ventures, I did not want to enter a new or sunrise sector. I wanted to enter a sector that would offer strong cash flows or be able to command a premium in capital markets.

These ideas guided me to focus on the infrastructure sector. A growing country was welcoming the participation of the private sector in infrastructure. While I was considering various options, we were introduced to infrastructure entrepreneur Arun Lakhani by my long-time friend Nitin Gadkari, now the Union minister of road transport, in 2006. Lakhani had won a concession for building highways in Maharashtra under the public-private partnership (PPP) model.

This was the opportunity that I had been waiting for. We teamed up with Lakhani and invested in his company. Our journey in the infrastructure sector began with this venture.

Later, we participated on our own in various projects under the PPP model. We have built more than 5,000 lane kilometres of roads under the BOT (build, operate and transfer) model with state and Central government licences. Many other companies that bid for similar projects are struggling under high debt as they could not sustain their aggressive promises. I am proud to say that we managed to avoid the fate of such companies by working hard and delivering results. We are now in the process of acquiring the rights to develop another 5,000 lane kilometre road assets, making us one of the largest road development companies in India.

Our journey in the infrastructure sector has not been limited to roads and highways. We have also entered other verticals like power transmission and renewable energy generation (solar, wind and hydro). Our group is expanding in environmental infrastructure such as converting municipal solid waste into energy, sewage water treatment and related activities.

We are attempting to pioneer a new concept in this field. We want to become a one-stop shop for utilities solutions. I would like our group to offer integrated solutions to needs like power, water, gas and broadband supply. Here we would be delivering directly to the consumers. This has positioned us uniquely as a 360-degree solution provider to support the Smart City campaign of Prime Minister Narendra Modi. We are in forefront of this game-changing initiative of the Government of India.

I will not be surprised if our media and infrastructure businesses find themselves competing to be the number one wealth creator of our group. The order book of the infrastructure company could exceed $20 billion in the next three to five years.

WE STARTED THE infrastructure business with a small team of twelve members who were determined to achieve the vision of ensuring smarter road connectivity. To learn the intricacies of the business, we started with an investment of $7.8 million in Malegaon-Manmad-Kopargaon roads project spearheaded by the Maharashtra government.

During 2007-08, we won four more state highway projects from Madhya Pradesh Road Development Corporation Limited (MPRDC). These helped us build our technological capabilities and a large portfolio size to attract talent and funding. Our endeavour was to build and provide state-of-the-art infrastructure assets and services in partnership with world class associates. In 2010, we won our first NHAI (National Highways Authority of India) contract—the Ahmedabad-Godhra project—at a cost of over $167 million. Yet another key milestone was the Sion-Panvel project, which saw successful completion in a short span of ten months despite many issues and challenges.

Today, we have completed around 2,100 lane kilometres and are currently constructing 2,500 lane kilometres in the states of Madhya Pradesh, Gujarat, Punjab, Haryana, Uttar Pradesh, Maharashtra and Tamil Nadu. I believe we are on the right track—ready for a focused and progressive journey in the infrastructure space.

The total investment in infrastructure is expected to be $1 trillion, with roads and bridge infrastructure accounting for $100 billion. The Government of India has earmarked $7.53 billion to develop a hundred smart cities across the country. It is encouraging to see an increase of $10.61 billion in allocation of funds for investments in infrastructure in the Union budget for 2015-16.

ELECTRICITY, WATER AND transport are the three most important pillars for the development of any region. My vision for mass urban utilities was to offer integrated services through a single conduit and one single bill for all utility services. In 2011-12, we launched Smart Utilities, wherein we could deliver integrated solutions through innovation and technology.

We started our utilities journey with a power distribution foray in Nagpur, with a franchisee project in contract with the Maharashtra State Electricity Distribution Company. We overcame the barriers we faced through our perseverance and were able to bring down the power distribution losses in Nagpur from 32 per cent to 15 per cent in a record time of just eighteen months. Thereafter, we repeated the same in Muzaffarpur, where we brought down the losses from 75 per cent to 21 per cent. After this, there was no looking back, and we won three more distribution franchises in Madhya Pradesh.

Our foray in water distribution started with a project awarded to us in Aurangabad. The idea was to enable equitable distribution of clean drinking water 24X7 to the citizens of Aurangabad and by far we have improved the situation significantly by plugging major leakages, which has resulted in a reduction in incidence of water-borne diseases.

We are currently the number one private power distribution franchise in India with a presence in five cities. We are also the largest public-private partnership operator in water distribution with projects in Aurangabad (Maharashtra), Bhagalpur (Bihar) and Tonk (Rajasthan). Our work in the municipal solid waste to energy management has also paid off and we are the country's top operator in this field as of now.

The Government of India has set an ambitious target of 175 GW green power production by 2022 and will cut the intensity of carbon emissions by up to 35 per cent and increase the share of non-fossil fuel based energy sources to 40 per cent by 2030. The Essel Group has made 12,000 MW of green energy commitment to the government wherein 7,500 MW of solar power, and 4,000 MW of wind energy projects will be undertaken over the next five years.

THE CONCEPT OF a Smart City came to me during my travels around the world. Whether it was Seattle, Barcelona, London, Chicago or Tokyo, what ruled the roost in these urban conglomerates was the definitive efficiency with which they were able to support rising populations, and the demand for and improved quality of lives among its inhabitants. Smart Cities as envisioned by us draw from the Vaastu Shastra and comprise the five components of nature or 'tatwas'—namely, the sun, air, sky, water and earth.

The pursuit of building Smart Cities is also being given emphasis by the new government. The ministry of urban development has defined Smart Cities as those that include the three pillars of competitiveness, sustainability and quality of life. By 2030, it is predicted that urban pockets will house over 70 per cent of the nation's population. This burgeoning population will be contributing around 75 per cent of the country's GDP. The preparation for such a future will require comprehensive development of physical, institutional, social and economic infrastructure of urban conglomerates.

In each endeavour that I have undertaken, each new goal I have set for myself since 2007, the building of Smart Cities has held centre stage. Today, we possess the capability of managing all components of Smart Cities. We are the only player in our sector to have proposed a governance framework and financial model to the Government of India for the building of these cities.

I feel proud that within a short span of nine years, Essel Infra is amongst the top five players in the country with an order book value $4.4 billion, having a presence in fifty-two cities across fifteen states of India. My desire, as stated earlier, is that this business should be valued in excess of $20 billion in three to five years' time.

And with these lines I embark on many more exciting journeys to come...

'What is needed is a vision, a mission and a burning desire to "make it happen"—regardless of wherever you are.'

EPILOGUE

LIFE IS ALWAYS full of crises, hope, challenges and enjoyment. How you look at it depends on your attitude.

Growing up in a reasonably well-off family, perhaps I would have been satisfied with what we had. But the turning point in my life came when our family lost its money and was almost bankrupt.

This reversal of fortune when I was just seventeen sparked in me a determination to fight against adverse circumstances. I had nothing with me except my will to overcome our family's financial crisis and win against all odds.

This determination took me from a small town to the megapolis of Delhi; I arrived with only Rs 17 in my pocket. Today, I am successful and wealthy by Indian standards.

But am I satisfied and happy? In some ways, yes. But in others, no.

The determination to do more, to learn more, has stayed with me. Somehow, my successes have not stopped me from seeking new challenges.

One aspect of my yearning has changed, though: I have decided that whatever business I do (even existing ones) will be with a social objective. An example is my ambition to develop Smart Cities. I want to create cities where health, education and economic opportunities support the aspiration of citizens.

I also want to develop Smart Villages that prevent needless migration and allow rural India to modernize in a sustainable manner. I want to create a model that can rejuvenate village life in a wholesome way, with support from private and government bodies.

I also keenly support the 'Make in India' campaign so that it creates long-term employment for millions while yielding growth for our group. After all, can't a profit-making company that helps the community be called a social enterprise?

In addition to being a motivator, I will set up a fund to help not only start-ups but young entrepreneurs facing a difficult phase in their business.

Being in media with a large reach of one billion viewers/ readers, we will soon launch a platform between change seekers and change agents, hence becoming a change maker. Our target is to reach 3 billion viewers in the next seven years, which will catapult us into a much higher profit orbit.

No country can develop without a political will. I want to do my bit to create an enviornment for Indian polity to become more development-oriented. It needs to come out of the socialist mindset of making the masses dependent on subsidies. A difference can be made by creating public awareness.

Though I personally believe that for 1.25 billion of us, being Indian is our first religion, I don't think it is reflected in our daily lives. I want to unite and help my community of 'Vaishyas' and urge them to be 'Indians' first and 'Vaishyas' second.

The other areas of philanthropy, cultural and social activities, will remain my focus through the 'DSC Foundation', albeit in a more structured manner. I do have to fulfil my duties towards my family, some of which remain incomplete.

In conclusion, I remain a hungry person who 'does not want to die', who wishes to be remembered after my current journey ends some day.

APPENDICES

Indian Media & Entertainment Industry Size (2014)

		Rs bn	USD bn
🖵	TV	475	7.4
🖶	Print	263	4.1
🎥	Films	126	2.0
📻	Radio	17	0.3
🎵	Music	10	0.2
🛆	Out of Home	22	0.3
🐴	Animation/VFX	45	0.7
🎮	Gaming	24	0.4
🖥	Digital Advertising	43	0.7
	Total	**1,026**	**16.0**

(Exchange Rate: 1 USD = INR64)
Source: FICCI-KPMG Indian Media and Entertainment Industry Report 2015

Essel / Zee Group Revenues 2015

Listed Companies	$ Mn
Zee Entertainment Enterprises Ltd	786
Zee Media Corporation Ltd	85
Dish TV India Ltd	438
Siti Cable Network Ltd	144
Zee Learn Ltd	20
Shirpur Gold Refinery Ltd	496
Unlisted Companies	
Essel Infra + Essel utilities	476
Pan India Network Ltd (Playwin)	491
Veria International Ltd	35
Essel Finance	2
Total	**2,973**

INDEX

Aap ki Adalat, 164, 197
Advani, L.K., 199
advertising revenues, importance of, 139, 140, 145, 152–53, 154, 174
Afro-Asian Satellite Communications, 248
Agrani (satellite project), 234, 238, 248, 250, 253, 254, 255
Ambani, Dhirubhai, 202, 204, 210
Ambani, Mukesh, 204, 210
Ambanis, enmity over news reporting, 203
Amit (son), 225, 233, 234, 235, 236
arms, business of importing, 103
Ashok (brother), 49, 73, 112, 210, 225, 234, 243
Asia Today Ltd, 136, 139–41, 159, 174, 178, 179, 181, 247
AsiaSat satellite, 124–25, 127, 133, 136–37, 175, 176

banks, experience with, 50, 53, 69–70, 73–74, 90–91, 93, 141
Basu, Rathikant (DD & News Corp), 180, 182, 187, 188, 190
Bhan, Chander, 24–25
Bhattacharya, Bhaskar, 97
Bhattacharya, Ranjan, 200
Bhave, C.B. (SEBI), 209

Birla, Aditya, support from, 186
Brahmachari, Swami Dhirendra, & rice deal, 79, 88, 97–99, 102, 103–04
downfall of, 98–99, 100–01
broadcasting sector, entry into, 97, 122–25, 126–28, 147, 152–53, 165, 169, 185, 238, 264

Chandra, Subhash
business challenges, resolving
befriending officials, 41–42, 53, 59
favours extended, 76
overcoming rules, 57–58, 84, 166
business failures, 37, 118
business success, 6, 31, 37, 61, 72, 73, 76, 104, 112, 122, 131, 151, 152, 153, 155, 156, 18 5, 187, 194, 231, 271
cheated by
Ghisa Ram, 71
Lal, C., 36, 37
Om Prakash, 58–60
Rakesh Gupta, 52, 71
creative business schemes, 28–29, 44–45
family commitments, 28, 37, 48–49

foreign trip (first), 84–86
friends, valuable help from, 6,
    11, 30, 35, 39, 41, 42, 51, 53,
    54, 55, 63, 69, 74, 76, 77, 79,
    83, 90, 96, 108, 109, 115, 121,
    133, 197, 204, 235, 249
learning experience, 71–72
marriage of, 45–48
marriage, views on, 28, 236
personal traits, 3, 4, 72, 122, 237,
    238, 272
    anger, 38, 47, 65, 128, 130,
        134, 194, 216, 233, 240–41
values of, 236
Carnegie, Andrew (News Corp),
    175, 176
Central Bureau of Investigation
    (CBI), 73–74, 77, 263
Chan brothers—Robert and
    George (Star), 127
Chanana, Anil (contractor), 40,
    41, 76, 77, 78, 80, 82–83, 109
Cheated on techno-economic
    feasibility report, 123, 248
Chisholm, Sam (News Corp), 165,
    175, 177, 179
Chopra, B.R., 36, 66–67, 141
CNN, 121, 128–29, 168

Dadaji, 4–5, 6–8, 11, 12–13, 14–
    15, 16, 17, 19–21, 22–23, 28, 37,
    38, 43–44, 46–48, 59, 72, 122,
    233, 234–35, 236, 238, 243, 244
Dalmiya, Jagmohan, & telecast
    rights, 212–18
Damodaran, M. (SEBI), 208
Deora, Murli, 157, 204
Desai, Manubhai (friend), 138
Dhar, Vijay, 80–81, 85, 87, 98–99,
    101, 102, 107, 251–52

Dhawan, R.K., 67, 98
donations to political parties, 204,
    244
Doordarshan (DD), 97, 121, 123–
    24, 126, 127, 133, 141–42, 148,
    149, 164, 165, 168, 169, 180, 247

Enforcement Directorate raids
    Zee TV, 172, 173
ESPN-Star Sports, 213, 214, 215,
    216
Essel Infra, 269
Essel Packaging Ltd., 85, 105–08,
    112, 116, 135, 192, 201
Essel World (amusement park),
    114–18, 120, 122, 238
Exportkhleb, 81, 82, 94–96,
    101
exports of basmati rice to USSR,
    78–83
    challenges, 88–89
    consignment approval, 94–96
    dealing with contacts, 78–82,
        97–101
    finance, 83, 89–91
    shipping issues, 92–93

family worked as a team, 1–2,
    11, 29–30, 49, 55, 56, 58, 60–
    61, 67, 68, 70, 85–86, 158, 172,
    229
family feud and settlement,
    158–59, 191–93, 227–28, 242–
    44
father, 3, 4, 5–6, 7, 8–10, 16–17,
    28, 30, 46, 48, 49, 191–93, 225,
    236, 244
First Break All the Rules by
    Marcus Buckingham and
    Curt Coffman, 241

forefathers and their business,
1–3, 4–8, 11–16

Gadkari, Nitin, 265
Gandhi, Indira, 76, 77, 80, 99–
100, 103–04
Gandhi, Rahul, 264
Gandhi, Rajiv, 79–80, 81, 99, 100,
101, 102–03, 108, 115, 138–39,
189
Gandhi, Sanjay, 75, 76, 77, 81, 101
Gandhi, Sonia, 263, 264
George, Vincent, 80, 103
Ghisa Ram (uncle), 7, 21–22, 71
Goenka, Ganpat Ram, 4
Goenka, Gopi Ram (relative and
friend), helped by, 4, 83, 88–89,
95
Goenka, Ram Gopal, 4
Goenkaji, Acharya Satya Narain,
222–24, 225
differences with, 228–30
problems with Pagoda building,
226–28
gold-refining project, 205, 235
Goldsmith, Sir James, 139, 176
Goyal, Sandeep (CEO), 213–14
Greenwood, David, 250
Grover, Subhash, 69, 83, 85, 93
Guha, Pradeep (CEO), 194–96, 242
Gujral, I.K., & DTH in India, 183–
84
Gupta, Kishan (helpful friend), 54
Gupta, Purushottam, 57, 62, 75,
84, 85
Gupta, Rakesh, let down by, 51–
53, 71, 72

Hamarey PMji, 172–73, 197
Hansaria, B.D., tricked by, 36

Helpline, 167
Hisar, 1, 6, 10, 12, 13, 26, 27, 28,
35, 37, 38, 42, 48, 49, 54, 63,
121, 151
helpful contacts of, 31, 39, 42,
54, 79, 121
Hughes Communications, 136,
248, 250–55

Indian Cricket League (ICL), 218–
21
Indian Space Research
Organisation (ISRO), 182, 184
infrastructure sector, focus on,
265–69
power transmission, 266
Smart utilities, 266, 267–68
state highway projects, 265–66,
267
water distribution, 268
Irani, Smriti, 149
Isaac, Ranjan, 126–27, 136, 140, 150
Iyer, T.N.V., 188

Jagan Nath, see Dadaji
Jain TV, 119
Jain, Dr J.K., 119
Jaitley, Arun, 215, 216
Jaitley, Jaya, 196
Jawahar (brother), 6, 29, 48, 49,
56, 58, 66, 83, 85, 88, 99, 156,
158, 168, 169, 171, 172, 193, 243,
244, 261–62
Jindal, Naveen, 257, 258–63
Jindal, Vijay (CEO; created rift
among brothers), 156–57, 158–
59, 193, 212
Johnson, Michael, duped by, 136–
37, 247–48, 249
joint venture with

News Corp, 165, 175–78, 181, 185, 186, 189, 190, 213
Star TV, 127–28, 130, 136, 178, 181, 211, 213

Kapoor, Atul (Star), 127–28, 129
Kaun Banega Crorepati (Star TV show), 160, 190
Keni, Nitin (Zee), 157, 161
Kesri, Sitaram, 102, 138
Kharakediwala, Radha Kishan, let down by, 43
Khariawala, Hari Ram, helpful advice from, 27
Kotak, Naresh, 92
Krishnan, Ananda, 182–83, 184
Kurien, Ashok, 126–27, 129, 132, 139, 140, 142, 148, 154, 157

Lakhani, Arun, 265–66
Lal, Bhajan, 79
Lal, C., let down by, 36, 65–71, 72, 77
Lal, Chaggan, 57, 58, 59–60
Lamina Packers, 62–63, 77
land in Bombay, experience of buying, 113–14
Laxman, Bangaru, 198, 199
Laxmi (brother), 6, 10, 21–22, 49, 55, 58, 148, 193, 261
Li Ka-shing, 125, 136–37, 175, 179
Li Richard (son of Li Ka-shing), 125, 129, 130–37, 148, 151–52, 175, 178, 179

Malhotra, Iqbal (News Corp), 176
Maliwal, Gopal (Zee), 154
Mandi Adampur, 1–2, 3, 4, 5, 6, 9, 10, 12, 15, 27, 41, 79, 244

Manion, David (Star), 127–28, 129, 132
Marwah, Ranjan, 129, 133, 135
McFarlen, Ted, 25, 252
MIPCOM (global trade exhibition), 161
Mishra, Brajesh, 200
Mittal, Rajinder (helpful friend), 79, 80, 83, 88, 98
Mohan, Narendra, 185
Moss, Steve (Star), 129, 132, 134–35, 248, 249, 250
movie rights, 127–28, 145–47
Mukerjea, Peter, 190, 195
Murdoch, Rupert, 175, 176, 179, 180–83, 184, 187, 188, 189, 190, 212, 247
Muthreja, Ravi, 259

Naik, Ram, 114–15, 116
Nehru, Arun, 80, 98, 102
Nepal, 120–21
News Corp, 165, 175–81, 185, 186, 189, 190, 213
Non-resident Indian (NRI), 123, 136, 151, 171

Packer, Kerry, 139, 176, 220
Pakistan, 166–67, 218
Parekh, Ketan, 190, 206, 207, 209
Parle Group, 111
partners for the TV project, 126, 133, 136
Patel, Mukesh (friend), 201–05, 206
Patel, Praful, 202
Pawar, Sharad, 115, 116, 117, 197, 217–18
Pingle, Dr (FCI), 52, 56, 63
Pooja (daughter), 235–36

not reach the mobile phone inside a building. That would have made the phone useless for people. I did not understand for a long while why Ashutosh Garg was singing Hughes's tune.

We had yet to finalize our paperwork with financial institutions such as IDBI. Our corporate rivals did not want this project to succeed. They tried to influence the government to not allow state-owned financial institutions to lend to us. Meanwhile, another satellite phone project, Iridium, backed by Motorola, had also launched and Indian institutions such as IFCI had invested in it.

Whenever Agrani had a meeting, they would talk about Iridium. They were betting on the Iridium project and not on ours. Getting the finance together for the project was becoming tough.

One fine morning Hughes sent us a notice cancelling the deal. They put us on notice by labelling us as defaulters.

I wondered what had gone wrong. The delayed payments for the satellite didn't seem to be the reason for the notice. There was something else behind the notice.

My feeling is that as Hughes had been hungry for the deal, it had underestimated the cost of building the satellite. Instead of making a profit, they would apparently have lost about $100 million on the deal. They had made a big miscalculation in the numbers. Instead of asking for a renegotiation of the deal, they wanted to wriggle out of it, and used delayed payment as an excuse. They had tried to use Ashutosh Garg to dilute the specifications but we did not agree to it. My British chief engineer on the project gave me honest advice that stopped me from diluting the specifications. I owe him a lot.

By then I had paid Hughes about $80 million. And before that I had spent more than $40 million on the project report and $10 million in running the company, thus sinking a total of $130 million in all.

Before Hughes cancelled the agreement, another hurdle had come our way—India had conducted nuclear tests. This led to sanctions by the US and its allies against India and companies from India. This resulted in us not being able to keep the schedule of payment to Hughes. After the sanctions came into effect, US companies could not accept money from Indian companies. We had raised this with Hughes, and they said they understood the situation. In fact they also lobbied with their government to permit them to sell the satellite to us.

The US government in its sanctions had said that their companies could not provide high-tech projects to India, which impacted our Agrani satellite as well. This was a serious issue. I contacted the then telecom minister, Ramvilas Paswan, who was in London. I met him there and requested him to write a letter to exempt our satellite from the banned list. Paswan sent the letter to the foreign minister, who in turn sent it to our ambassador in the US. The ambassador lobbied with the US government on this issue. With some persuasion we managed to get an exemption from the US government on our satellite delivery. But this did not solve our problem, as Hughes was not in a mood to deliver. Despite the effort, they sent a cancellation notice to wriggle out of a loss-making deal for them.

We filed a case against Hughes, asking for the appointment of an arbitrator by both sides, which they were refusing. We wanted them to return the payment made by us till then. The court ordered Hughes to accept our plea and appointed an arbitrator. After more than a year of fighting we won the arbitration but the arbitrator ordered that the money be returned without interest. That was a substantial loss for us. Hughes paid $80 million (without interest) but I had already lost $50 million out of the total spend of $130 million.

This was a setback to my plans. If the satellite had been built to the specifications, we could have become Asia's biggest telecom company.

Prakash, Om, duped by, 58–60
Prakash, Sanjeev (ANI), 161–62
Prasad, Inder, (granduncle), 4, 7, 11, 16, 17
programmes, creating, 141–43
public-private partnership (PPP) model, 265–66
Punit (son), 196, 225, 233, 234–35, 236, 241, 256

Rajju Bhaiya, 187
Ramji, Mangat (neighbour from Hisar), helped by, 39, 42
Rao, Narasimha, 170, 171–73, 182
Rashtriya Swayamsevak Sangh (RSS), 10, 11, 186, 187, 189, 200
Rastogi, Suresh, 53
Rathi, Mr, deceived by, 204–05
Ray, Siddharth (Star), 176
regulations of government, 114–15, 120, 128, 165, 181, 186, 188, 208
circumvent the, 165

Sachdeva, Gulshan (friend), 121, 122–23, 142, 163
Saha, A.C., 193
Samtani, Karuna, 142, 154, 156, 174
Sanyal, B.K. (CEO), 155, 156
satellite mobile phone business, 255–56
SatelliteTelevision for Asian Region (Star) Ltd, 124–25, 127–28, 129–37, 140, 160, 169, 175–81, 184, 187, 188, 190, 195
Sayagyi U Ba Khin, 223
Screwvala, Ronnie, 142
SEBI sends show-cause to Zee Telefilms, 208–09

Shah, Dhirubhai, court case filed by, 146–47
Sharma, Mr, 75–76
Sharma, Rajat (Zee), 162–67, 197
fall out with, 167–68
Shiv Sena, 201–02
Shukla, Rajiv, 215–16
Shyam, Radhey (schoolmate), 17, 18
Singh, Amar, 199, 210
Singh, Chaudhary Aman (FCI), 31–34
Singh, Digvijay (Zee), 154, 156, 157, 174
Singh, R.K. (CEO), 160, 190, 193
Singh, Ram (friend), 17–18
Sinha, Yashwant, 199
Siti Cable, 179, 193, 194
Smart City, 266, 267, 269, 271
Smart Villages, 271–72
Srinivasan, N., 215, 216
Star TV, violation of agreement by, 181
stock market crash, 207–09
stocks options, 160
Subramaniam, T.S.R., 183–84
Sudharshan, K.S., 187
Suraj Lamp, 67–69, 71
Sushila (wife), 46, 48, 225, 236

Tandon, Ashok, 199
Tanna Exports, 94–95, 101
Tanna, Tulsi, 87, 88–90
Tehelka, 198, 199, 208
telecast rights for
World Cup, 211–14
Indian cricket, 214–17
Thawani, Harish, 212, 213

Tripathi, Mr, 75, 76, 77, 81, 101
TV channels lost money and
    folded up, 187, 196

Union Carbide, 51, 52, 55

Vakil, Hitesh, 157
Vipassana, 131, 222–31, 240
Vishwanathan, C.S., 240–41

Zee Business, 169, 259
Zee channels, other, 156, 179, 194,
    211, 214, 235
Zee Jagran, 194

Zee News, 165–67, 168–69, 193
    negative side-effects with
        Ambanis, 203, 209
        Narasimha Rao, 171, 172
        Naveen Jindal, 257, 262
        Vajpayee, 197, 199
Zee News, coal scandal on, 257–
    63
Zee Telefilms Ltd, 140, 159–60,
    174, 180, 192
Zee TV & News Corp deal, 175–
    81, 184, 187, 188, 189
Zee TV, 136, 159, 161, 166, 193,
    194, 195
    benefits of, 167, 170

ABOUT THE AUTHORS

Subhash Chandra is the promoter of Essel/Zee Group of companies, which is a major player in the fields of media and entertainment, packaging, technology, infrastructure and education.

Pranjal Sharma has been in print, digital and TV media for twenty-five years. He has led teams at India Today Group and CNBC Network 18, and was founding executive editor of Bloomberg TV in India. He currently writes for *Businessworld* and hosts a show for Zee Business.